My Book

This book belongs to

Name: _____

Copy right © 2019 MATH-KNOTS LLC

All rights reserved, no part of this publication may be reproduced, stored in any system or transmitted in any form, or by any means, electronic, mechanical, photocopying, recording, or otherwise without the written permission of MATH-KNOTS LLC.

Cover Design by :
Gowri Vemuri

First Edition :
April, 2020

Second Edition :
January, 2021

Author :
Gowri Vemuri

Editor :
Ritvik Pothapragada

Questions: mathknots.help@gmail.com

Dedication

This book is dedicated to:
My Mom, who is my best critic, guide and supporter.
To what I am today, and what I am going to become tomorrow,
is all because of your blessings, unconditional affection and support.

This book is dedicated to the
strongest women of my life,
my dearest mom
and
to all those moms in this universe.

G.V.

BASIC ALGEBRA 1

INDEX

Notes	9 - 12
Writing verbal expressions Algebraically	13 - 58
Writing Numerical and Algebraic expressions in words	59 - 69
Writing Algebraic equations in words	70 - 76
Writing Algebraic inequalities in words	77 - 83
Order of operations : Numeric expressions	84 - 88
Order of operations : Algebraic expressions (substituting given values)	89 - 95
Simplifying Expressions	96 - 105
One step equations	106 - 118
Two step equations	119 - 124
Inequalities	125 - 144
Answer keys	145 - 206

Integers notes

Integers can be represented on a number line as below

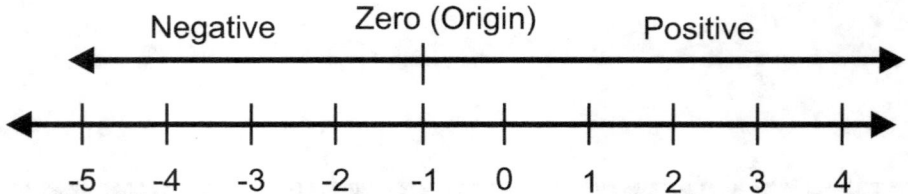

Absolute value :

The number line can be used to find the absolute value. The absolute value of an integer is the distance the number is from zero on the number line.

The absolute value of 2 is 2. Using the number line , 2 is a distance of 2 to the right of zero. The absolute value of -2 is also 2. Again using the number line , the distance from -2 to zero is 2. A measure of distance is always positive.

The symbol for absolute value of any number , x , is | x |.

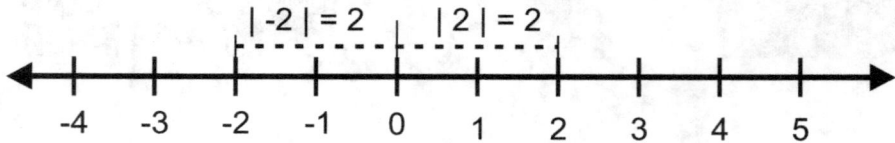

Opposite integers :

The opposite of an integer is the number that is at the same distance from zero in the opposite direction. Every integer has an opposite value, but the opposite of zero is itself.

The opposite of -4 is 4 because it is located the same distance from zero as 4 is , but in opposite direction.

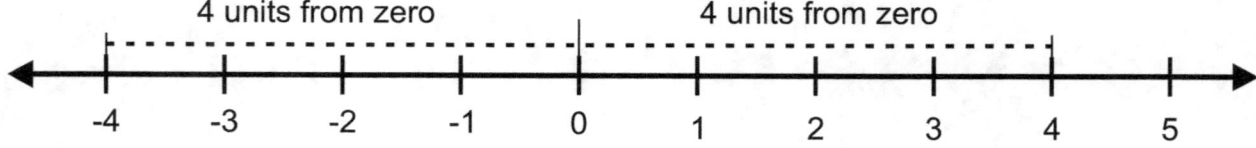

BASIC ALGEBRA 1

Adding integers using a number line :

The number line is visual representation to understand the addition of positive and negative numbers. Start with the one value on the number line, then add the second value. If the second value (that is added) is positive, we move to the right that many spaces.

If the second value (that is added) is negative, we move to the left that many spaces.
The value where we land on the number line is the solution for the addition of two integers.

Example 1 : (-4) + (5) = 1
Start at the first number, -4, and travel 5 units to the right.

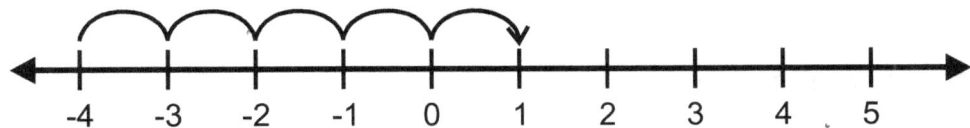

Example 2 : (5) + (-7) = -2
Start at the first number, 5, and travel 7 units to the left.

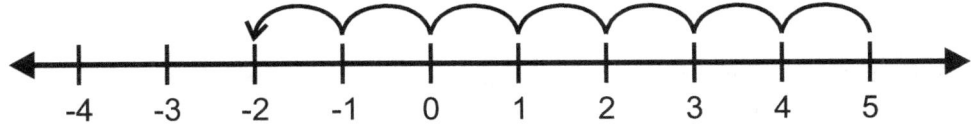

Adding integers using the rules :

Rules for adding integers :

If the signs are the same, add their absolute values, and keep the common sign.

If the signs are different, find the difference between the absolute values of the two numbers, and keep the sign of the number with the greater numerical value.

To the Tune of "Row Your Boat"

Same signs add and keep
Different signs subtract
Keep the sign of the greater digits
then you'll be exact

BASIC ALGEBRA 1

Notes

Subtacting integers using a number line :

A number line is helpful in understanding subtraction of positive and negative values. Start with the first value on the number line, then subtract the second value. If the second value (that is subtracted) is positive, we move to the left that many spaces.

If the second value (that is subtracted) is negative, we move to the right that many spaces. This is because subtraction a negative is the same as adding.
The value where we end on the number line is the answer.

Example 3 : (-2) + (5) = 3
Start at the first number, -2, and travel 5 units to the right.

Subtacting integers using the rules :

Every subtraction problem can be written as an additional problem. When we subtract two integers, just <u>ADD THE OPPOSITE.</u> Subtracting a positive is the same as adding a negative. Subtracting a negative is the same as adding a positive.

Multiplying Integers :

Multiplying integers is same as multiplying whole numbers, except we must keep track of the signs associated to the numbers.

To multiply signed integers, always multiply the absolute values and use these rules to determine the sign of the product value

When we multiply two integers with the same signs, the result is always a positive value.

Positive number X Positive number = Positive number

Negative number X Negative number = Positive number

When we multiply two integers with different signs, the result is always a negative value.

Positive number X Negative number = Negative number

Negative number X Positive number = Negative number

Positive X Positive :   7 X 6 = 42   negative X negative : -7 X -6 = 42

Positive X negative :   7 X -6 = -42   negative X Positive : -7 X 6 = -42

BASIC ALGEBRA 1

Notes

Dividing Integers :

Division of integers is similar to the division of whole numbers, except we must keep track of the signs associated.

To divide signed integers, we must always divide the absolute values and use the below rules to find the quotient value.

When we divide two integers with the same signs, the result is always a positive value.

$$\text{Positive} \div \text{Positive} = \text{Positive}$$

$$\text{Negative} \div \text{Negative} = \text{Positive}$$

When we divide two integers with opposite signs, the result is always a negative value.

$$\text{Positive} \div \text{Negative} = \text{Negative}$$

$$\text{Negative} \div \text{Positive} = \text{Negative}$$

Example 4 :

Positive ÷ Positive :	81 ÷ 9 = 9		Positive ÷ negative :	81 ÷ -9 = -9
negative ÷ negative :	-81 ÷ -9 = 9		negative ÷ Positive :	-81 ÷ 9 = -9

Golden Rules of Integers :

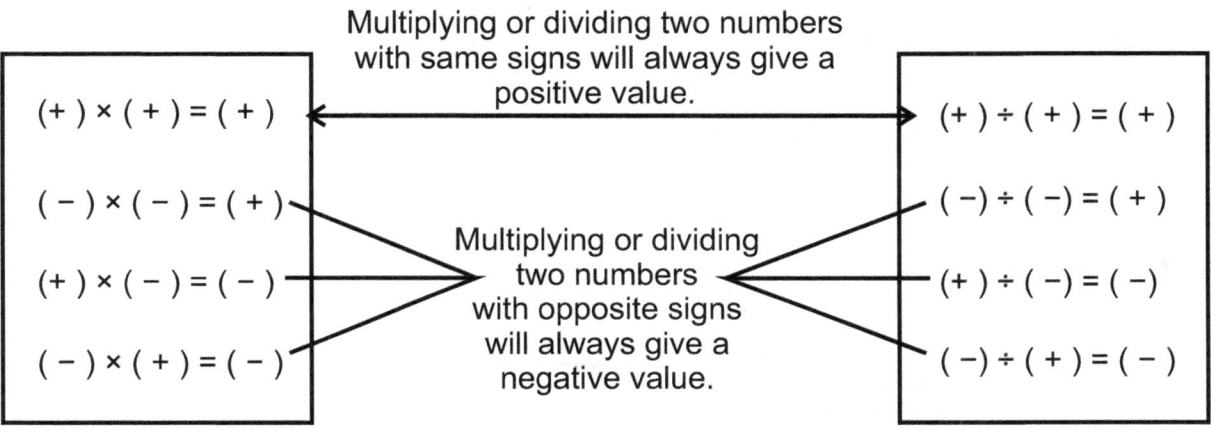

Introduction of Exponents

Whenever any integer, let us say x, is added n times, the result obtained will be equal to n times x i.e. nx. But in case, if the integer x is multiplied for n times, the result obtained will be equal to x^n (which is called as exponential form). The problems relating to these will be studied under 'Exponents'. We shall look at the rules/properties pertaining to these exponential numbers in this chapter.

Rational Exponents and Radicals

If 'a' is any real number and 'n' is a positive integer, then the product a × a × a × ---- n times is represented by the notation a^n. This notation is referred to as exponential form. In the above notation, a is called the base and n is called the power or exponent or index (plural of index is indices). a^n is read as 'n^{th} power of a' or 'a to the power n'.

Example 1: 6 × 6 × 6 × 6 × 6 × 6 × 6 can be written as 6^7. Here 6 is called as base and 7 is called as index (or exponent).

For a non-zero rational number 'a' with a negative integral exponent 'm' the following result can be observed.

$a^m = a^{-n} = a^{-1} \times a^{-1} \times a^{-1} \times a^{-1} \times ---- \times --- $ n times

$= \dfrac{1}{a} \times \dfrac{1}{a} \times \dfrac{1}{a} \times \dfrac{1}{a} ---- $ n times $= \left(\dfrac{1}{a}\right)^n = \dfrac{1}{a^n}$

Example 2: $6^{-3} = \left(\dfrac{1}{6}\right)^3$

If a is a positive rational number and n = p/q is a positive rational exponent, then we can define $a^{p/q}$ in two ways.

(1) $a^{\frac{p}{q}}$ is the qth root of a^p, i.e. $a^{\frac{p}{q}} = \left(a^p\right)^{\frac{1}{q}}$

(2) $a^{\frac{p}{q}}$ is the pth power of qth root of a, i.e. $a^{\frac{p}{q}} = \left(a^{\frac{1}{q}}\right)^p$.

Parts of the exponent:

This is read as "Seven to the fourth power"

BASIC ALGEBRA 1

Notes

Laws of indices:

For all real numbers a and b and all rational numbers m and n, we have

(i) $a^m \times a^n = a^{m+n}$

Example 3 : (1) $2^3 \times 2^6 = 2^{3+6} = 2^9$

(2) $\left(\dfrac{5}{6}\right)^4 \times \left(\dfrac{5}{6}\right)^5 = \left(\dfrac{5}{6}\right)^{4+5} = \left(\dfrac{5}{6}\right)^9$

(3) $5^{2/3} \times 5^{4/3} = 5^{(2/3 + 4/3)} = 5^{6/3} = 5^2$

(4) $2^3 \times 2^4 \times 2^5 \times 2^8 = 2^{(3+4+5+8)} = 2^{20}$.

(5) $(\sqrt{7})^3 \times (\sqrt{7})^{\frac{5}{2}} = (\sqrt{7})^{3+\frac{5}{2}} = (\sqrt{7})^{\frac{11}{2}}$

(ii) $a^m \div a^n = a^{m-n}, a \neq 0$

Examples 4 : (a) $7^8 \div 7^3 = 7^{8-3} = 7^5$

(b) $\left(\dfrac{7}{3}\right)^9 \div \left(\dfrac{7}{3}\right)^5 = \left(\dfrac{7}{3}\right)^{9-5} = \left(\dfrac{7}{3}\right)^4$

(c) $9^{\frac{2}{3}} \div 9^{\frac{1}{6}} = 9^{\left(\frac{2}{3} - \frac{1}{6}\right)} = 9^{\left(\frac{4-1}{6}\right)} = 9^{\frac{3}{6}} = 9^{\frac{1}{2}}$

(d) $\left(\dfrac{5}{7}\right)^{\frac{8}{9}} \div \left(\dfrac{5}{7}\right)^{\frac{1}{3}} = \left(\dfrac{5}{7}\right)^{\left(\frac{8}{9} - \frac{1}{3}\right)} = \left(\dfrac{5}{7}\right)^{\frac{8-3}{9}} = \left(\dfrac{5}{7}\right)^{\frac{5}{9}}$

Note: $a^n \div a^n = 1$
or $a^{n-n} = a^0 = 1$
$\therefore a^0 = 1, a \neq 0$

(iii) $(a^m)^n = a^{m \times n}$

Examples 5 :   (a) $(5^2)^3 = 5^{2 \times 3} = 5^6$

(b) $\left[\left(\dfrac{2}{3}\right)^4\right]^5 = \left(\dfrac{2}{3}\right)^{4 \times 5} = \left(\dfrac{2}{3}\right)^{20}$

(c) $\left[\left(\dfrac{5}{7}\right)^{\frac{2}{3}}\right]^{\frac{9}{8}} = \left(\dfrac{5}{7}\right)^{\left(\frac{2}{3} \times \frac{9}{8}\right)} = \left(\dfrac{5}{7}\right)^{\frac{3}{4}}$

BASIC ALGEBRA 1

Notes

(iv) $\left(\dfrac{a}{b}\right)^n = \dfrac{a^n}{b^n}$

    Example 6 : $\left(\dfrac{4}{5}\right)^7 = \dfrac{4^7}{5^7}$ **Note:**   Conversely we can write $\left(\dfrac{a^n}{b^n}\right) = \left(\dfrac{a}{b}\right)^n$

    Example 7 : $\dfrac{8}{27} = \dfrac{2^3}{3^3} = \left(\dfrac{2}{3}\right)^3$

(v) $(ab)^n = a^n \times b^n$

    Example 8 :   (a) $20)^5 = (4 \times 5)^5 = 4^5 \times 5^5$

                      (b) $(42)^7 = (2 \times 3 \times 7)^7 = 2^7 \times 3^7 \times 7^7$

Note: Conversely we can write $a^n \times b^n = (ab)^n$

Example 9 :   (a) $\left((5)^3\right)^2 = 5^{3 \times 2} = (5)^6 = 5 \times 5 \times 5 \times 5 \times 5 \times 5 = 15625$

                (b) $(2)^3 \times (2)^5 = (2)^{3+5} = (2)^8 = 2 \times 2 \times 2 \times 2 \times 2 \times 2 \times 2 \times 2 = 256$

                (c) $(7)^0 = 1$ $\boxed{\text{Any base value rise to the power zero is always equal to 1}}$

                (d) $(3)^{-4} = \dfrac{1}{(3)^4} = \dfrac{1}{3 \times 3 \times 3 \times 3} = \dfrac{1}{81}$

                (d) $\dfrac{(8)^7}{(8)^5} = (8)^{7-5} = (8)^2 = 64$

                (e) $\dfrac{(9)^4}{(9)^7} = (9)^{4-7} = (9)^{-3} = \dfrac{1}{(9)^3} = \dfrac{1}{9 \times 9 \times 9} = \dfrac{1}{729}$

                (f) $(2)^4 = 2 \times 2 \times 2 \times 2 = 16$ (g) $(-2)^4 = -2 \times -2 \times -2 \times -2 = 16$

                (h) $-(2)^4 = -(2 \times 2 \times 2 \times 2) = -16$ (i) $-(2)^3 = -(2 \times 2 \times 2) = -8$

                (j) $(-2)^3 = -2 \times -2 \times -2 = -8$ (k) $-(-2)^3 = -(-2 \times -2 \times -2) = -(-8) = 8$

Tip 1 : When the exponent is an even number the simplified value is always positive, when the base has a positive or negative value.

Tip 2 : When the exponent is an odd number the simplified value is always positive, when the base has a positive value.

Tip 3 : When the exponent is an odd number the simplified value is always negative, when the base has a negative value.

PEDMAS (or PEMDAS) rules :

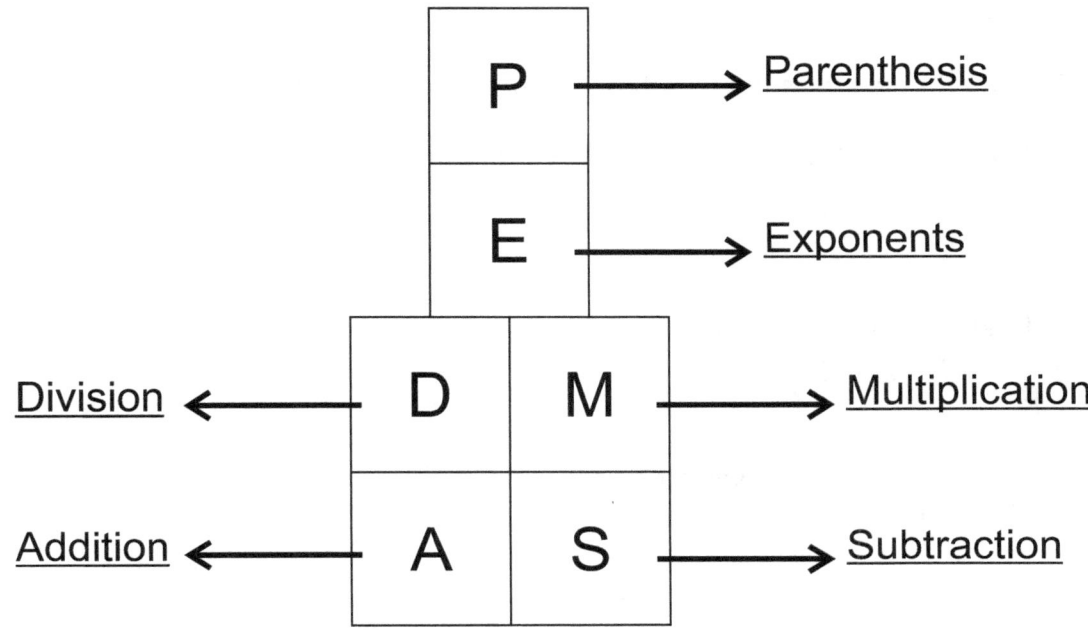

Steps to solve numerical or algebraic expressions and equations :

Step 1. Convert all division into a fraction (doing this way will avoid making silly mistakes).
Step 2. Solve the values within the parenthesis using PEDMAS rules.
   Within the parenthesis follow step 3 to step 8 as applicable.
Step 3. Check for anymore simplifications within the parenthesis using PEDMAS rules.
Step 4. Check for any exponential values if exists simplify otherwise proceed to step 5.
Step 5. Check for division and simplify as needed, if no division values exists proceed to step 6.
Step 6. Check for multiplications and simplify, if no multiplication exists proceed to step 7.
Step 7. Check for addition and simplify. Remember that a number should be considered along with its sign associated before the number. If no addition exists proceed to step 8.
Step 8. Check for subtraction and simplify. Remember that a number should be considered along with its sign associated before the number.

Rules of addition and subtraction of integers should be applied as needed for step 7 and 8.
Rules of multiplication and division of integers should be applied as needed in
step 1 through step 6.

BASIC ALGEBRA 1

Order of operations:

Example 10:

$(12 + 4)^2 \div 8 - 3^2 \times 2 + 5 \times 2 - 10$

Using PEDMAS rules we need to solve the numerical expressions.

Step 1: (Convert $\div$ to a fraction and rewrite the numerical expression)

$\dfrac{(12 + 4)^2}{8} - 3^2 \times 2 + 5 \times 2 - 10$

Step 2: (First solve the parenthesis)

$\dfrac{(16)^2}{8} - 3^2 \times 2 + 5 \times 2 - 10$

Step 3: Check for any more parenthesis to simplify if none are there then check for exponent values and simplify them.

$\dfrac{256}{8} - 9 \times 2 + 5 \times 2 - 10$

Step 4: Multiply as needed in the numerical expression

$\dfrac{256}{8} - 18 + 10 - 10$

Step 5: Divide as needed in the numerical expression

$32 - 18 + 10 - 10$

Step 6: Simplify using addition and subtraction rules and pay attention to positive and negative signs.

$32 - 18 + 10 - 10 = 14 + 0 = 14$

BASIC ALGEBRA 1

Notes

Example 11 :

Simplify $x^2 + y(2x + y - 1) - 5$; where $x = 5$ and $y = 3$

Substitute the given values of x and y in the expression $x^2 + y(2x + y - 1) - 5$ and then solve.

Using PEDMAS rules we need to solve the numerical expressions.

Step 1 : (Convert ÷ to a fraction and rewrite the numerical expression if division operation is given in the expression otherwise proceed to step 2)

$5^2 + 3(2(5) + 3 - 1) - 5$

Step 2 : (First solve the parenthesis)

$5^2 + 3(2(5) + 3 - 1) - 5 = 5^2 + 3(10 + 3 - 1) - 5$

Step 3 : Check for any more parenthesis to simplify if none are there then check for exponent values and simplify them.

$5^2 + 3(10 + 3 - 1) - 5 = 5^2 + 3(13 - 1) - 5 = 5^2 + 3(12) - 5$
$5^2 + 3(12) - 5 = 25 + 3(12) - 5$

Step 4 : Multiply as needed in the numerical expression

$25 + 3(12) - 5 = 25 + 36 - 5$

Step 5 : Divide as needed in the numerical expression

$25 + 36 - 5$ (there are no division operation in this example)

Step 6 : Simplify using addition and subtraction rules and pay attention to positive and negative signs.

$25 + 36 - 5 = 61 - 5 = 56$

BASIC ALGEBRA 1

Notes

Basic Algebra notes

Expressions	Numerical expression	Algebraic/Variable expression
Expressions will not include equal sign	12^2 $8 + 10$ $4^2 - 16 + 9$	$6x + 17$ $5a^2 - 2b$
Equation	Numerical equation	Algebraic/variable equation
Must include an equal sign; One side is equal to the other side	$9 + 8 = 17$ $5^2 - 11 + 2 = 16$	$6x + 5 = 35$ (x=__?__) $9x^2 = 225$ (x=__?__)

Solving Equations :

To solve an equation first add the like terms (if any) and then isolate the variable on to one side of the equation. To solve one - step equations, do the inverse operation to find the value of the variable.

Remember : Always perform the same operation on both sides of the equation to maintain the balance.
Inverse operation for addition is subtraction and vice versa.
Inverse operation for multiplication is division and vice versa.

Polynomial	Highest degree	1st Name by degree	2nd Name by number of terms
8	0	Constant	monomial
$9x + 8$	1	Linear	binomial
$-2x^2 + 7x - 11$	2	Quadratic	trinomial
$8x^3$	3	Cubic	monomial
$2x^4 + x^3$	4	Quartic	binomial
$3x^5 + 5x^3 + 12$	5	Quintic	trinomial
$2x^6 + x^5 - 3x^4 - 21$	6	6th degree	Polynomial with 4 terms
$5x^7 - x^5 + 4x^4 - 7x^2 + x - 3$	7	7th degree	Polynomial with 6 terms

BASIC ALGEBRA 1

Notes

A Polynomial in standard form is always written with terms in sequential order according to their highest degree of exponents (higher exponents to lower exponents).

Like Terms :

Two or more terms are said to be alike if they have the same variable and the same degree. Coefficients of like terms are not necessarily be same.

An expression is in its simplest form when

1. All like terms are combined.
2. All parentheses are opened and simplified.

Like Terms can combined by adding or subtracting their coefficients (pay attention to the positive and negative signs of the coefficient and apply rules of adding integers)

Example 1 : $-2x + 5x + 7 = 3x + 7$

Note : -2x and 5x are like terms and can be combined using rules of integers

Example 2 : $-11y + 5 + 8y - 7 = -3y - 2$

Note : -11y and 8y are like terms and can be combined using rules of integers. 5 and -7 are like terms and can be combined using rules of integers

Example 3 : $-12a - 5a + 8 - 3 = -17a + 5$

Note : -12a and -5a are like terms and can be combined using rules of integers. 8 and -3 are like terms and can be combined using rules of integers

Example 4 : $-5b + 7 - 3b + 2a - a + 10 = -8b + a + 17$

Note : -5b and -3b are like terms and can be combined using rules of integers. 2a and -a are like terms and can be combined using rules of integers. 7 and 10 are like terms and can be combined using rules of integers.

Combining like terms on the opposite side of the equal sign :

When the like terms are on opposite sides, we have to combine like terms by using the inverse operation and by undoing the equation.

Example 5 : $-2x + 5 = -7x$

$$-2x + 5 = -7x$$
$$+7x \quad\quad +7x$$
$$-2x + 7x + 5 = -7x + 7x$$
$$5x + 5 = 0$$

Solving equations using the distributive property :

The number in front of the parentheses needs to be multiplied with every term within the parentheses. After the distribution and opening up the parentheses, combine like terms and solve.

Distributing with the negative sign :

Remember to apply the integer rules of positive and negative numbers while distributing.

$$+ \times + = +$$
$$- \times - = +$$
$$- \times + = -$$
$$+ \times - = -$$

Example 6 : (a) $2(5x + 7) = 2(5x) + 2(7) = 10x + 14$
(b) $-7(3a + 8) = (-7)(3a) + (-7)(8) = -21a + (-56) = -21a - 56$
(c) $3(-5b - 2) = (3)(-5b) - (3)(2) = -15b - 6$
(d) $6(-4a + 5) = (6)(-4a) + (6)(5) = -24a + 30$
(e) $4(2a - 8) = (4)(2a) - (4)(8) = 8a - 32$
(f) $-5(a - 7) = (-5)(a) - (-5)(7) = -5a - (-35) = -5a + 35$
(g) $-9(-2a + 10) = (-9)(-2a) + (-9)(10) = 18a + (-90) = 18a - 90$
(h) $-8(-5a - 6) = (-8)(-5a) - (-8)(6) = 40a - (-48) = 40a + 40a$
(i) $-(a + 7) = -a - 7$
(j) $-(x - 5) = (-1)(x) - (-1)(5) = -x - (-5) = -x + 5$
(k) $-(-a - b) = (-1)(-a) - (-1)(b) = a - (-b) = a + b$
(l) $-(-a + 2b) = (-1)(-a) + (-1)(2b) = a + (-2b) = a - 2b$

BASIC ALGEBRA 1

Notes

Example 7:
$$2x + 3 = x + 7$$
$$2x + 3 = x + 7$$
$$\underline{-x\ -3\ \ -x\ -3}$$
$$x + 0 = 0 + 4$$
$$x = 4$$

Inverse operation for addition is subtraction

Example 8:
$$7x + 5 = -3x + 25$$
$$7x + 5 = -3x + 25$$
$$\underline{3x\ -5\ \ \ 3x\ \ -5}$$
$$10x + 0 = 0 + 20$$
$$10x = 20$$
$$\frac{10x}{10} = \frac{20}{10}^{2}$$
$$\boxed{x = 2}$$

Inverse operation for addition is subtraction and vice versa

Inverse operation for multiplication is division

Example 9:
$$\frac{2x}{5} + 5 = 15$$

$$\frac{2x}{5} + 5 = 15$$
$$\underline{\ \ \ \ \ -5\ \ \ -5}$$
$$\frac{2x}{5} + 0 = 10$$

$$\frac{2x}{5} = 10$$

$$5 \cdot \frac{2x}{5} = 5 \cdot 10$$

$$\frac{2x}{2} = \frac{50}{2}^{25}$$

$$\boxed{x = 25}$$

Inverse operation for addition is subtraction and vice versa

Inverse operation for division is multiplication

Inverse operation for multiplication is division

BASIC ALGEBRA 1

Notes

Inequality :

An inequality is a relation between two expressions that are not equal. As a mathematical statement an inequality states one side of the equation is less than, less than or equal to or greater than or greater than equal to the other side.

If the inequality has **less than** or **greater than** symbol,
1.   The graph starts with the open circle.
2.   For less than the graphing line goes toward the left.
3.   For greater than the graphing line goes toward the right.

If the inequality has **less than or equal to** or **greater than or equal** to symbol,
1.   The graph starts with the closed circle.
2.   For less than or equal to the graphing line goes toward the left.
3.   For greater than or equal to the graphing line goes toward the right.

Inequality statement	Inequality verbal expression	Inequality graph
x > -3 or -3 < x	x is greater than -3	
x < 3 or 3 > x	x is less than 3	
x >= -1 or -1 <= x	x is greater than or equal to -1	
x <= 1 or 1 <= x	x is less than or equal to 1	

Basic inequalities :

Solving inequalities is same as solving for an equation except for one special rule.

BASIC ALGEBRA 1

Notes

Example 10 : x + 9 > 11
Step 1 (subtract 9 from both sides) : x + 9 - 9 > 11 - 9
Step 2 (combine like terms) : x > 2

Example 11 : 2x + 5 > 10
Step 1 (subtract 5 from both sides) : 2x + 5 - 5 > 10 - 5
Step 2 (combine like terms) : 2x > 5
Step 3 (divide both sides by the coefficient of x which is 2) : $\frac{\cancel{2}x}{\cancel{2}} > \frac{5}{2}$

Step 4 (simplify both sides as needed) : $x > \frac{5}{2}$

Example 12 : 5x - 1 > 9
Step 1 (add 1 to both sides) : 5x - 1 + 1 > 9 + 1
Step 2 (combine like terms) : 5x > 10
Step 3 (divide both sides by the coefficient of x which is 5) : $\frac{\cancel{5}x}{\cancel{5}} > \frac{10}{5}$

Step 4 (simplify both sides as needed) : x > 2

Example 13 : 2x - 8 > -11
Step 1 (add 8 to both sides) : 2x - 8 + 8 > -11 + 8
Step 2 (combine like terms) : 2x > -3
Step 3 (divide both sides by the coefficient of x which is 2) : $\frac{\cancel{2}x}{\cancel{2}} > \frac{-3}{2}$

Step 4 (simplify both sides as needed) : $x > \frac{-3}{2}$

Example 14 : -4x + 7 > 10
Step 1 (subtract 7 from both sides) : -4x + 7 - 7 > 10 - 7
Step 2 (combine like terms) : -4x > 3
Step 3 (divide both sides by the coefficient of x which is -4) : $\frac{\cancel{-4}x}{\cancel{-4}} < \frac{3}{-4}$

> When an inequality is multiplied or divided with negative number, the inequality changes to the opposite

Step 4 (simplify both sides as needed) : $x < \frac{3}{-4}$

$$x < \frac{-3}{4}$$

BASIC ALGEBRA 1

Notes

Example 15 : -5x - 1 < -11
Step 1 (add 1 to both sides) : -5x - 1 + 1 < -11 + 1
Step 2 (combine like terms) : -5x < -10
Step 3 (divide both sides by the coefficient of x which is -5) : $\frac{-5x}{-5} > \frac{-10}{-5}$

> When an inequality is multiplied or divided with negative number, the inequality changes to the opposite

Step 4 (simplify both sides as needed) : x > 2

Example 16 : -8x + 3 < -30
Step 1 (subtract 3 from both sides) : -8x + 3 - 3 < -30 - 3
Step 2 (combine like terms) : -8x < -33
Step 3 (divide both sides by the coefficient of x which is -8) : $\frac{-8x}{-8} > \frac{-33}{-8}$

Step 4 (simplify both sides as needed) : $x > \frac{-33}{-8}$

$$x > \frac{33}{8}$$

$$x > 4\frac{1}{8}$$

Transform each of the below verbal expressions into an algebraic expression.

(1) The sum of a number and 12

(2) 7 plus a number

(3) 9 more than a number

(4) A number increased by 5

(5) n more than 12

(6) A number increased by 11

(7) The sum of 9 and a number

(8) A number plus 10

(9) The sum of 2 and a number

(10) 12 plus 6

BASIC ALGEBRA 1

Transform each of the below verbal expressions into an algebraic expression.

(11) A number plus 12

(12) 3 plus a number

(13) A number increased by 12

(14) A number increased by 8

(15) 6 more than 6

(16) The sum of 11 and 12

(17) The sum of 7 and 11

(18) The sum of a number and 11

(19) n more than 8

(20) A number increased by 6

Transform each of the below verbal expressions into an algebraic expression.

(21) 11 decreased by 9

(22) 21 less than 25

(23) The difference of a number and 15

(24) 17 less than w

(25) 9 less than n

(26) The difference of 23 and a number

(27) 12 decreased by 3

(28) r less than 20

(29) n less than 15

(30) A number decreased by 15

Transform each of the below verbal expressions into an algebraic expression.

(31) The difference of a number and 6

(32) 20 decreased by 19

(33) 13 decreased by 9

(34) The difference of 27 and 14

(35) 3 less than p

(36) The difference of a number and 7

(37) The difference of a number and 25

(38) 13 minus 7

(39) 10 less than 12

(40) 12 minus 10

Transform each of the below verbal expressions into an algebraic expression.

(41) The product of a number and 9

(42) Twice a number

(43) The product of 8 and 5

(44) The product of a number and 7

(45) 6 times a number

(46) Twice 11

(47) A number times 6

(48) The product of a number and 11

(49) 11 times a number

(50) 2 times a number

Transform each of the below verbal expressions into an algebraic expression.

(51) The product of 8 and 8

(52) Twice 12

(53) The product of 5 and 10

(54) Twice 5

(55) 8 times a number

(56) The product of 8 and a number

(57) The product of 2 and 10

(58) Twice 10

(59) The product of 2 and a number

(60) 3 times 8

BASIC ALGEBRA 1

Basic Math

Transform each of the below verbal expressions into an algebraic expression.

(61) 16 divided by 2

(62) The quotient of 24 and 4

(63) A number divided by 2

(64) The quotient of a number and 4

(65) 6 divided by a number

(66) The quotient of 14 and 2

(67) Half of 14

(68) The quotient of 33 and 3

(69) Half of 6

(70) A number divided by 4

Transform each of the below verbal expressions into an algebraic expression.

(71) Half of 8

(72) A number divided by 3

(73) 36 divided by 4

(74) 6 divided by 2

(75) 33 divided by a number

(76) 20 divided by 4

(77) The quotient of 25 and 5

(78) Half of a number

(79) A number divided by 8

(80) Half of 16

BASIC ALGEBRA 1

Transform each of the below verbal expressions into an algebraic expression.

(81) The 2nd power of 5

(82) 3 squared

(83) 9 squared

(84) p squared

(85) The 4th power of 3

(86) The 2nd power of 8

(87) 6 squared

(88) u to the 3rd

(89) The 2nd power of 2

(90) 5 cubed

Transform each of the below verbal expressions into an algebraic expression.

(91) The y power of 10

(92) 4 cubed

(93) The y power of 14

(94) The n power of 11

(95) 3 to the y

(96) The n power of 9

(97) 4 cubed

(98) The 3rd power of 3

(99) 5 cubed

(100) n cubed

BASIC ALGEBRA 1

Transform each of the below verbal expressions into an algebraic expression.

(101) The sum of a number and 8 is 45

(102) 8 more than a number is 39

(103) 10 more than a number is equal to 24

(104) The sum of a number and 9 is equal to 7

(105) A number plus 11 is equal to 34

(106) 11 more than a number is 14

(107) A number plus 6 is 27

(108) A number increased by 10 is equal to 48

(109) The sum of a number and 12 is equal to 16

(110) The sum of a number and 10 is equal to 19

BASIC ALGEBRA 1

Basic Math

Transform each of the below verbal expressions into an algebraic expression.

(111) A number plus 9 is equal to 33

(112) A number increased by 5 is 24

(113) A number increased by 8 is 12

(114) The sum of a number and 5 is 29

(115) 9 more than a number is 24

(116) A number increased by 7 is 18

(117) A number plus 7 is 31

(118) A number plus 5 is 17

(119) A number increased by 9 is 30

(120) 7 more than a number is 18

Transform each of the below verbal expressions into an algebraic expression.

(121) 12 less than n is equal to 8

(122) 9 less than d is 22

(123) The difference of a number and 10 is equal to 6

(124) A number minus 10 is 30

(125) The difference of a number and 14 is 24

(126) A number decreased by 13 is equal to 20

(127) A number minus 6 is 13

(128) A number minus 4 is 40

(129) A number decreased by 9 is equal to 36

Transform each of the below verbal expressions into an algebraic expression.

(130) A number decreased by 8 is equal to 48

(131) A number minus 14 is equal to 6

(132) A number minus 16 is 48

(133) 3 less than k is 33

(134) 8 less than m is 47

(135) A number minus 7 is 32

(136) The difference of a number and 4 is equal to 50

(137) 4 less than x is 18

Transform each of the below verbal expressions into an algebraic expression.

(138) A number decreased by 16 is equal to 27

(139) A number minus 8 is equal to 31

(140) 6 less than w is equal to 19

(141) A number times 12 is 48

(142) A number times 9 is 18

Transform each of the below verbal expressions into an algebraic expression.

(143)   The product of a number and 7 is equal to 37

(144)   A number times 5 is 42

(145)   A number times 8 is 44

(146)   The product of a number and 10 is 9

(147)   A number times 6 is 9

Transform each of the below verbal expressions into an algebraic expression.

(148)   The product of a number and 5 is equal to 30

(149)   Twice a number is 6

(150)   The product of a number and 6 is equal to 43

(151)   A number times 7 is 19

(152)   The product of a number and 11 is 15

Transform each of the below verbal expressions into an algebraic expression.

(153)   A number times 10 is 46

(154)   The product of a number and 8 is equal to 21

(155)   The product of a number and 12 is equal to 40

(156)   The quotient of a number and 3 is equal to 46

(157)   A number divided by 5 is equal to 30

Transform each of the below verbal expressions into an algebraic expression.

(158)   A number divided by 4 is equal to 32

(159)   Half of a number is 16

(160)   A number divided by 7 is equal to 20

(161)   A number divided by 6 is 20

(162)   A number divided by 2 is equal to 47

Transform each of the below verbal expressions into an algebraic expression.

(163) The quotient of a number and 7 is 32

(164) The quotient of a number and 6 is 16

(165) The quotient of a number and 4 is 44

(166) A number divided by 8 is equal to 44

(167) The quotient of a number and 2 is 47

Transform each of the below verbal expressions into an algebraic expression.

(168) The quotient of a number and 8 is 36

(169) The quotient of a number and 5 is equal to 36

(170) A number divided by 3 is 42

(171) The 4th power of q is 40

(172) The v power of 4 is 44

Transform each of the below verbal expressions into an algebraic expression.

(173) 7 to the n is equal to 33

(174) The 6th power of x is equal to 28

(175) 5 to the m is 15

(176) The 3rd power of t is 35

(177) The 5th power of a is equal to 47

Transform each of the below verbal expressions into an algebraic expression.

(178) t to the 9th is 20

(179) The 7th power of x is equal to 22

(180) 15 to the n is 50

(181) 9 to the n is 30

(182) A number cubed is 42

(183) The 9th power of n is equal to 10

(184) A number squared is 27

(185) The 8th power of x is 20

(186) The n power of 12 is equal to 30

(187) n to the 6th is equal to 50

Transform each of the below verbal expressions into an algebraic expression.

(188) n to the 7th is equal to 42

(189) The k power of 5 is 46

(190) y to the 4th is 46

(191) A number increased by 7 is greater than or equal to 23

(192) A number decreased by 10 is greater than or equal to 13

Transform each of the below verbal expressions into an algebraic expression.

(193)   12 more than a number is less than or equal to 45

(194)   A number minus 5 is less than 28

(195)   A number plus 12 is less than or equal to 50

(196)   The difference of a number and 16 is greater than or equal to 46

(197)   The difference of a number and 9 is less than 22

Transform each of the below verbal expressions into an algebraic expression.

(198)   A number plus 12 is greater than 48

(199)   17 less than x is less than or equal to 15

(200)   A number decreased by 9 is greater than 11

(201)   A number plus 10 is less than or equal to 19

(202)   The sum of a number and 8 is less than or equal to 13

Transform each of the below verbal expressions into an algebraic expression.

(203) The difference of a number and 20 is less than 34

(204) A number decreased by 9 is greater than or equal to 15

(205) The difference of a number and 11 is greater than or equal to 24

(206) The difference of a number and 15 is less than 18

(207) 10 more than a number is greater than 20

Transform each of the below verbal expressions into an algebraic expression.

(208) 5 more than a number is greater than or equal to 36

(209) The difference of a number and 6 is greater than 35

(210) A number increased by 5 is less than 12

(211) A number plus 11 is less than or equal to 33

(212) A number decreased by 9 is less than or equal to 31

Transform each of the below verbal expressions into an algebraic expression.

(213) The sum of a number and 9 is less than or equal to 21

(214) A number plus 8 is less than or equal to 50

(215) A number minus 23 is less than 32

(216) A number minus 20 is greater than or equal to 38

(217) The difference of a number and 8 is less than 16

Transform each of the below verbal expressions into an algebraic expression.

(218) A number increased by 11 is greater than or equal to 8

(219) The sum of a number and 9 is less than 10

(220) 3 less than w is greater than 12

(221) A number increased by 9 is less than 46

(222) A number plus 12 is greater than or equal to 20

Transform each of the below verbal expressions into an algebraic expression.

(223)   A number minus 15 is greater than or equal to 27

(224)   11 less than a is less than or equal to 15

(225)   A number increased by 5 is less than or equal to 18

(226)   5 more than a number is less than 21

(227)   The sum of a number and 10 is less than or equal to 14

Transform each of the below verbal expressions into an algebraic expression.

(228)   A number plus 8 is greater than or equal to 13

(229)   A number minus 12 is less than or equal to 29

(230)   The sum of a number and 11 is greater than 26

(231)   The quotient of a number and 8 is less than or equal to 10

(232)   A number divided by 6 is greater than or equal to 14

Transform each of the below verbal expressions into an algebraic expression.

(233) Twice a number is greater than or equal to 29

(234) A number cubed is greater than 6

(235) A number times 12 is greater than or equal to 24

(236) The product of a number and 8 is greater than 24

(237) A number divided by 7 is greater than or equal to 17

Transform each of the below verbal expressions into an algebraic expression.

(238)   The product of a number and 7 is less than or equal to 43

(239)   A number squared is less than or equal to 50

(240)   Half of a number is greater than or equal to 31

(241)   Half of a number is less than or equal to 11

(242)   The product of a number and 6 is less than or equal to 43

Transform each of the below verbal expressions into an algebraic expression.

(243) A number divided by 4 is greater than 39

(244) The product of a number and 6 is less than 16

(245) A number times 11 is less than 16

(246) Twice a number is less than 10

(247) Twice a number is less than or equal to 13

Transform each of the below verbal expressions into an algebraic expression.

(248) A number times 9 is less than or equal to 47

(249) A number times 7 is greater than or equal to 12

(250) The product of a number and 7 is greater than or equal to 26

(251) Half of a number is greater than 43

(252) A number divided by 2 is less than 40

Transform each of the below verbal expressions into an algebraic expression.

(253)   A number divided by 3 is less than 22

(254)   A number squared is less than 22

(255)   The quotient of a number and 3 is greater than 45

(256)   A number squared is greater than 17

(257)   A number divided by 8 is less than 48

Transform each of the below verbal expressions into an algebraic expression.

(258)   The product of a number and 12 is greater than 20

(259)   A number divided by 7 is less than or equal to 10

(260)   A number cubed is less than 48

(261)   The quotient of a number and 3 is less than 30

(262)   A number divided by 4 is less than 40

Transform each of the below verbal expressions into an algebraic expression.

(263) The quotient of a number and 4 is greater than 10

(264) A number divided by 2 is greater than or equal to 17

(265) The quotient of a number and 5 is less than or equal to 40

(266) A number divided by 6 is less than or equal to 41

(267) The quotient of a number and 6 is greater than 39

Transform each of the below verbal expressions into an algebraic expression.

(268) A number times 6 is greater than or equal to 19

(269) The quotient of a number and 4 is less than or equal to 39

(270) The quotient of a number and 5 is greater than or equal to 18

(271) A number squared is less than or equal to 17

(272) 9 to the p is less than 49

Transform each of the below verbal expressions into an algebraic expression.

(273) A number cubed is less than 41

(274) The 9th power of p is less than 45

(275) A number cubed is less than or equal to 28

(276) A number cubed is greater than 47

(277) v to the 10th is greater than 9

Transform each of the below verbal expressions into an algebraic expression.

(278)　The x power of 10 is less than or equal to 36

(279)　n to the 9th is greater than 21

(280)　A number cubed is greater than or equal to 49

(281)　The n power of 5 is less than 22

(282)　A number squared is greater than or equal to 15

Transform each of the below verbal expressions into an algebraic expression.

(283) The 8th power of x is greater than or equal to 39

(284) A number squared is less than 39

(285) n to the 6th is greater than or equal to 37

(286) The c power of 11 is greater than 49

(287) 6 to the n is greater than or equal to 27

Transform each of the below verbal expressions into an algebraic expression.

(288)   The t power of 15 is less than or equal to 7

(289)   n to the 2nd is greater than or equal to 33

(290)   15 to the w is less than 12

BASIC ALGEBRA 1

Transform each of the below numeric or algebraic expressions into a verbal expression.

(291) 30 − 19

(292) $n + 10$

(293) 8 + 6

(294) 16 − 4

(295) $n + 5$

(296) $11 + n$

(297) 12 + 9

(298) $11 + n$

(299) $n − 11$

(300) 8 + 11

BASIC ALGEBRA 1

Basic Math

Transform each of the below numeric or algebraic expressions into a verbal expression.

(301) $12 - 7$

(302) $5 + n$

(303) $23 - 4$

(304) $n - 8$

(305) $4 + n$

(306) $n + 7$

(307) $4 + n$

(308) $11 + 6$

(309) $8 + n$

(310) $n - 7$

BASIC ALGEBRA 1

Basic Math

Transform each of the below numeric or algebraic expressions into a verbal expression.

(311) $n + 11$

(312) $8 + n$

(313) $18 - 11$

(314) $n + 7$

(315) $9 + 9$

(316) $21 - 15$

(317) $8 + n$

(318) $c - 6$

(319) $10 - 5$

(320) $29 - 12$

BASIC ALGEBRA 1

Transform each of the below numeric or algebraic expressions into a verbal expression.

(321) $n + 9$

(322) $12 - r$

(323) $30 - 21$

(324) $n - 6$

(325) $15 - 4$

(326) $11 + 7$

(327) $n + 8$

(328) $n + 8$

(329) $10 + 12$

(330) $n - 12$

BASIC ALGEBRA 1

Transform each of the below numeric or algebraic expressions into a verbal expression.

(331) $4 \cdot 6$

(332) $\dfrac{40}{8}$

(333) $2 \cdot 6$

(334) $4n$

(335) $\dfrac{n}{2}$

(336) $n \cdot 10$

(337) $11 \cdot 6$

(338) $\dfrac{n}{6}$

(339) $\dfrac{77}{7}$

(340) $\dfrac{n}{3}$

BASIC ALGEBRA 1

Basic Math

Transform each of the below numeric or algebraic expressions into a verbal expression.

(341) $\dfrac{16}{2}$

(342) $2 \cdot 12$

(343) $\dfrac{20}{2}$

(344) $n \cdot 8$

(345) $\dfrac{14}{2}$

(346) $\dfrac{42}{6}$

(347) $\dfrac{12}{2}$

(348) $2n$

(349) $\dfrac{16}{8}$

(350) $\dfrac{n}{4}$

Transform each of the below numeric or algebraic expressions into a verbal expression.

(351) $4n$

(352) $\dfrac{n}{5}$

(353) $\dfrac{36}{n}$

(354) $n \cdot 6$

(355) $2 \cdot 3$

(356) $\dfrac{4}{n}$

(357) $2 \cdot 7$

(358) $2 \cdot 2$

(359) $\dfrac{10}{2}$

BASIC ALGEBRA 1

Transform each of the below numeric or algebraic expressions into a verbal expression.

(360) $2 \cdot 9$

(361) $\dfrac{77}{7}$

(362) $\dfrac{n}{6}$

(363) $\dfrac{n}{2}$

(364) $2 \cdot 10$

(365) $3n$

(366) $\dfrac{12}{n}$

(367) $8 \cdot 9$

(368) $\dfrac{8}{2}$

(369) $6 \cdot 8$

BASIC ALGEBRA 1

Transform each of the below numeric or algebraic expressions into a verbal expression.

(370) $n \cdot 5$

(371) $2 \cdot 8$

(372) $2 \cdot 12$

(373) $12n$

(374) $11 \cdot 9$

(375) $\dfrac{4}{2}$

(376) $n \cdot 11$

(377) $\dfrac{30}{n}$

(378) $\dfrac{60}{n}$

(379) $2n$

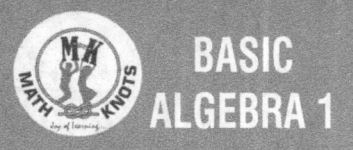

BASIC ALGEBRA 1

Transform each of the below numeric or algebraic expressions into a verbal expression.

(380) $2 \cdot 5$

(381) n^3

(382) n^2

(383) 6^2

(384) n^8

(385) 10^2

(386) 10^2

(387) 9^t

(388) 5^3

(389) 3^3

BASIC ALGEBRA 1

Transform each of the below numeric or algebraic expressions into a verbal expression.

(390) 5^2

(391) 4^c

(392) 5^3

(393) 7^2

(394) 4^2

(395) 5^2

(396) 6^2

(397) 2^3

(398) 12^n

(399) d^6

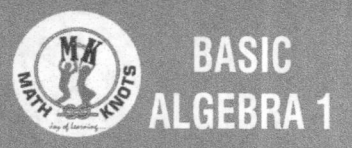

BASIC ALGEBRA 1

Basic Math

Transform each of the below numeric or algebraic expressions into a verbal expression.

(400) 3^3

(401) $n - 9 = 44$

(402) $n + 7 = 27$

(403) $n - 12 = 14$

(404) $u - 21 = 46$

(405) $n + 11 = 45$

(406) $n - 6 = 34$

(407) $n - 8 = 47$

(408) $n \cdot 11 = 37$

(409) $n \cdot 7 = 31$

BASIC ALGEBRA 1

Basic Math

Transform each of the below numeric or algebraic expressions into a verbal expression.

(410) $n - 6 = 50$

(411) $n - 20 = 46$

(412) $n - 5 = 18$

(413) $n \cdot 6 = 43$

(414) $n - 13 = 40$

(415) $n + 6 = 13$

(416) $n + 5 = 48$

(417) $n \cdot 9 = 34$

(418) $n - 3 = 11$

(419) $n - 9 = 22$

Transform each of the below numeric or algebraic expressions into a verbal expression.

(420) $n - 22 = 29$

(421) $n + 11 = 24$

(422) $n + 7 = 36$

(423) $n + 12 = 34$

(424) $2n = 19$

(425) $n - 12 = 18$

(426) $n - 3 = 46$

(427) $n \cdot 5 = 44$

(428) $n - 12 = 45$

(429) $n + 9 = 30$

BASIC ALGEBRA 1

Transform each of the below numeric or algebraic expressions into a verbal expression.

(430) $n - 17 = 33$

(431) $n - 4 = 37$

(432) $n \cdot 5 = 17$

(433) $n + 8 = 30$

(434) $n + 9 = 20$

(435) $n + 12 = 17$

(436) $n \cdot 7 = 22$

(437) $n - 3 = 29$

(438) $n + 5 = 43$

(439) $n \cdot 12 = 20$

Transform each of the below numeric or algebraic expressions into a verbal expression.

(440) $n - 22 = 34$

(441) $m - 14 = 8$

(442) $n - 14 = 8$

(443) $n + 11 = 16$

(444) $n - 7 = 49$

(445) $n - 20 = 8$

(446) $n + 12 = 15$

(447) $n + 9 = 33$

(448) $d - 12 = 12$

(449) $u - 4 = 13$

BASIC ALGEBRA 1

Basic Math

Transform each of the below numeric or algebraic expressions into a verbal expression.

(450) $n + 8 = 5$

(451) $w^{10} = 40$

(452) $\dfrac{n}{8} = 15$

(453) $12^m = 42$

(454) $13^n = 40$

(455) $8^x = 34$

(456) $4^x = 15$

(457) $n^3 = 49$

(458) $\dfrac{n}{7} = 33$

(459) $\dfrac{n}{2} = 25$

BASIC ALGEBRA 1

Transform each of the below numeric or algebraic expressions into a verbal expression.

(460) $n^2 = 46$

(461) $\dfrac{n}{5} = 27$

(462) $w^8 = 20$

(463) $\dfrac{x}{6} = 34$

(464) $p^2 = 7$

(465) $\dfrac{x}{2} = 6$

(466) $14^x = 46$

(467) $6^u = 27$

(468) $r^3 = 15$

(469) $14^m = 41$

BASIC ALGEBRA 1

Transform each of the below numeric or algebraic expressions into a verbal expression.

(470) $\dfrac{x}{4} = 36$

(471) $n \cdot 7 \geq 36$

(472) $\dfrac{n}{3} \leq 47$

(473) $r - 6 \leq 44$

(474) $\dfrac{x}{8} > 23$

(475) $n + 8 < 6$

(476) $\dfrac{n}{6} \leq 44$

(477) $n - 24 > 12$

(478) $r^2 \leq 26$

(479) $m + 10 \leq 29$

BASIC ALGEBRA 1

Transform each of the below numeric or algebraic expressions into a verbal expression.

(480) $r^5 \geq 40$

(481) $\dfrac{z}{6} \leq 13$

(482) $n + 5 > 39$

(483) $u^3 < 32$

(484) $\dfrac{x}{3} < 11$

(485) $2n > 28$

(486) $\dfrac{u}{8} \leq 22$

(487) $b + 9 > 20$

(488) $n - 16 < 32$

(489) $n + 8 \leq 49$

BASIC ALGEBRA 1

Transform each of the below numeric or algebraic expressions into a verbal expression.

(490) $y - 19 \geq 15$

(491) $n - 10 \geq 45$

(492) $d + 7 \geq 8$

(493) $c^8 \geq 49$

(494) $2b < 14$

(495) $\dfrac{z}{7} \geq 44$

(496) $z - 6 > 24$

(497) $n \cdot 8 > 5$

(498) $n^2 < 23$

(499) $n + 12 \leq 14$

Transform each of the below numeric or algebraic expressions into a verbal expression.

(500) $\dfrac{n}{2} \leq 33$

(501) $n^2 > 40$

(502) $\dfrac{n}{7} \geq 40$

(503) $n - 18 \leq 27$

(504) $n^6 > 12$

(505) $n \cdot 12 < 8$

(506) $n - 9 \geq 30$

(507) $n + 9 > 5$

(508) $n \cdot 5 \leq 48$

(509) $c - 9 \geq 13$

Transform each of the below numeric or algebraic expressions into a verbal expression.

(510) $\dfrac{n}{2} < 10$

(511) $y \cdot 6 < 32$

(512) $6^r < 7$

(513) $q - 5 \leq 22$

(514) $n \cdot 12 > 24$

(515) $p - 20 > 13$

(516) $n + 10 \geq 21$

(517) $9^x < 37$

(518) $n \cdot 5 \leq 19$

(519) $\dfrac{c}{7} > 30$

BASIC ALGEBRA 1

Transform each of the below numeric or algebraic expressions into a verbal expression.

(520) $y - 8 > 6$

(521) $2n \geq 7$

(522) $x \cdot 11 \leq 20$

(523) $x \cdot 10 < 49$

(524) $7^u > 5$

(525) $n + 6 \geq 38$

(526) $b \cdot 11 \geq 37$

(527) $\dfrac{u}{5} < 11$

(528) $m^3 > 10$

(529) $y - 7 < 37$

BASIC ALGEBRA 1

Transform each of the below numeric or algebraic expressions into a verbal expression.

(530) $r - 9 \geq 42$

BASIC ALGEBRA 1

Basic Math

Evaluate each of the below numerical expressions.

(531) $8 \div (4 - (2 + 4 - 4))$

(532) $4 - 3 + (10 - 4) \div 3$

(533) $(((3)(3))(2)) \div 3$

(534) $5 - 4 \div 2 + 2^2$

(535) $((4)(2)) \div (1 - 1 + 2)$

(536) $(2)(6) - 6 - (6 - 5)$

(537) $(3)(2) - (1 + 2) \div 3$

(538) $2 \div (5 - (4 - 3) - 2)$

(539) $3 + 4 + 2 - 3 - 3$

(540) $((5)(2)) \div 5 + 6$

BASIC ALGEBRA 1

Basic Math

Evaluate each of the below numerical expressions.

(541) $2 + 4 + (3)(2) - 2$

(542) $((2)(5))(6) - (4)(4)$

(543) $(9 + 8 - (3 + 2)) \div 2$

(544) $(6 - (1 + 2 - 1)) \div 4$

(545) $(5)(10 \div ((2)(2) - 2))$

(546) $6 + 6 - 10 \div (2 + 3)$

(547) $(6 + (5 + 5)(2))(2)$

(548) $(6)(1^2) + 18 \div 3$

(549) $(4)(5) - (6 + 5 - 1)$

(550) $(2 - 1)((4 + 1) \div 5)$

BASIC ALGEBRA 1

Basic Math

Evaluate each of the below numerical expressions.

(551) $(6 \div 3)((7 + 3) \div 2)$

(552) $((2 + 3)(3)) \div 5 - 2$

(553) $(4)((5)(5 - 3) - 5)$

(554) $(4 - 2)(5^2) - 4$

(555) $((5)(3)) \div (5 + 4 - 4)$

(556) $4 + 12 - (44 + 5) \div 7$

(557) $11 - 4 + 32 \div (9 - 1)$

(558) $(7 - 4 + 4 + 5) \div 4$

(559) $((20 - 9)(3)) \div (7 + 4)$

(560) $((4 + 16)(2)) \div 4 + 7$

Evaluate each of the below numerical expressions.

(561) $3 - (7 - 5) + 14 - 13$

(562) $5 + 9 + (2)(10 + 8)$

(563) $10 + (8)(14 - 9 + 1)$

(564) $(5 + 25) \div ((7 - 6)(10))$

(565) $11 - 14 \div 7 - 8 + 1$

(566) $3 + 15 - (1 + 45 \div 3)$

(567) $(9 - (12 - 10))(12 + 4)$

(568) $(27 + 3) \div (11 - 6 - 2)$

(569) $(((13)(3))(2)) \div (15 - 9)$

(570) $8 + 2 + 7 - (13 - 10)$

Evaluate each of the below numerical expressions.

(571) $3 + 8 + 14 + 10 - 2$

(572) $(15)(13 - 6 - 1) - 2$

(573) $(4)(12) - 20 \div (9 + 1)$

(574) $11 - 2 + 4 - 1 + 3$

(575) $(3 + 15 - 7)(1 + 5)$

(576) $4 + 15 - ((33)(3)) \div 11$

(577) $(((5^2)(2)) \div 5)(12)$

(578) $4 \div (6 - 2 - (7 - 5))$

(579) $40 \div (2 - 1 + 12 - 9)$

(580) $12 + 13 + (4)(15) - 6$

Evaluate each of the below expressions by using the given values.

(581) $j - h \div 3$; where $h = 3$, and $j = 7$

(582) $q + m - m$; where $m = 1$, and $q = 1$

(583) $(zy)^2$; where $y = 4$, and $z = 2$

(584) $5ba$; where $a = 5$, and $b = 2$

(585) $(m)(p \div 2)$; where $m = 7$, and $p = 10$

(586) $5 + zy$; where $y = 5$, and $z = 2$

(587) $x + y \div 4$; where $x = 7$, and $y = 8$

(588) $m - (n - n)$; where $m = 8$, and $n = 1$

(589) $y + x - 3$; where $x = 9$, and $y = 2$

(590) $x + y - y$; where $x = 4$, and $y = 2$

Evaluate each of the below expressions by using the given values.

(591) $a + b^2$; where $a = 5$, and $b = 2$

(592) $(y + x) \div 6$; where $x = 4$, and $y = 2$

(593) $y + x + y$; where $x = 10$, and $y = 2$

(594) $(p + q)^2$; where $p = 2$, and $q = 1$

(595) $(y - x)^2$; where $x = 4$, and $y = 7$

(596) $j^2 h$; where $h = 4$, and $j = 5$

(597) $k - (h - k)$; where $h = 7$, and $k = 5$

(598) $b - b + a$; where $a = 2$, and $b = 4$

(599) $y + y - z$; where $y = 10$, and $z = 1$

(600) $(m)(m - n)$; where $m = 10$, and $n = 8$

BASIC ALGEBRA 1

Basic Math

Evaluate each of the below expressions by using the given values.

(601) $m - (p - q^2)$; where $m = 10$, $p = 7$, and $q = 1$

(602) $y + (8 + x) \div 4$; where $x = 8$, and $y = 7$

(603) $9 - (b - b) - a$; where $a = 7$, and $b = 7$

(604) $(y + x - y) \div 5$; where $x = 5$, and $y = 12$

(605) $y + x + x + 11$; where $x = 13$, and $y = 5$

(606) $q - (r \div 4)^2$; where $q = 2$, and $r = 4$

(607) $x + z + 7 - x$; where $x = 12$, and $z = 9$

(608) $(p)(r + r - 2)$; where $p = 4$, and $r = 9$

(609) $(x + y) \div 4 + y$; where $x = 10$, and $y = 6$

(610) $(m)(p \div 2 + m)$; where $m = 6$, and $p = 14$

Evaluate each of the below expressions by using the given values.

(611) $n + 2m - p$; where $m = 11, n = 4,$ and $p = 4$

(612) $q - (12 - q + p)$; where $p = 3,$ and $q = 10$

(613) $mn - 14 \div 2$; where $m = 11,$ and $n = 9$

(614) $x + z - y + z$; where $x = 3, y = 3,$ and $z = 7$

(615) $r + (q)(15 \div 3)$; where $q = 4,$ and $r = 11$

(616) $h - (7 - j + j + 4)$; where $h = 14$, and $j = 4$

(617) $(x + 7 - (y - 3)) \div 4$; where $x = 18$, and $y = 12$

(618) $((20)(k - (k - h))) \div 4$; where $h = 16$, and $k = 19$

Evaluate each of the below expressions by using the given values.

(619) $6 + (a - b) \div 3 + b$; where $a = 14$, and $b = 11$

(620) $6x - x - y - x$; where $x = 17$, and $y = 1$

(621) $y - (16 + x - y) + x$; where $x = 9$, and $y = 17$

(622) $n - (n - (m - (m - n)))$; where $m = 17$, and $n = 10$

(623) $(n)(n + 5) - m^2$; where $m = 1$, and $n = 3$

Evaluate each of the below expressions by using the given values.

(624) $(y)(x+5) - (x+y)$; where $x = 8$, and $y = 10$

(625) $y + x + x - (y - y)$; where $x = 1$, and $y = 9$

(626) $12 + 8 - j + h \div 2$; where $h = 14$, and $j = 8$

(627) $nm - (7 + 13 - n)$; where $m = 9$, and $n = 10$

(628) $((y)(11 + x - x)) \div 4$; where $x = 10$, and $y = 8$

Evaluate each of the below expressions by using the given values.

(629) $y \div 3 - (9 - x^2)$; where $x = 3$, and $y = 15$

(630) $q - (q - (p - (p - q)))$; where $p = 5$, and $q = 5$

Simplify each of the below expressions.

(631) $3n - 2n$

(632) $a + 7 + 1 + 4a$

(633) $6x + 6 - 9x - 7$

(634) $2n + 2 - 3n$

(635) $7x - 5x$

(636) $10 + 9r + 1 - 10r$

(637) $9r - 5r$

(638) $7b - b$

(639) $-5 + 8x + 4$

(640) $8x + 2 - 8x$

Simplify each of the below expressions.

(641) $3p + 4p$

(642) $-7x - 5x$

(643) $-7n - 7n$

(644) $-3n - 3 - 1$

(645) $1 + 8n + n - 2$

(646) $-x + 9x$

(647) $1 + 2x + 2$

(648) $7x + 10x$

(649) $-9n - n$

(650) $2x + 3 - 3 - 9x$

Simplify each of the below expressions.

(651) $3x + x$

(652) $-b + 10b$

(653) $-3b - 8 + 5b$

(654) $-7n + 2n$

(655) $-9x - 8x$

(656) $-11k(-2k - 8)$

(657) $3(-1 - 9m)$

(658) $-2m(m - 1)$

(659) $-3b(b - 10)$

(660) $7(10 - 4v)$

BASIC ALGEBRA 1

Basic Math

Simplify each of the below expressions.

(661) $-9(-7x + 11)$

(662) $-7(12 - 11k)$

(663) $-5(4p - 5)$

(664) $-3x(10x + 2)$

(665) $4(11v + 3)$

(666) $-6k(5k + 11)$

(667) $-9a(1 + 9a)$

(668) $12(9x + 7)$

(669) $10(-2n - 3)$

(670) $-5p(p - 1)$

Simplify each of the below expressions.

(671) $-11(1-9k)$

(672) $-(10r+6)$

(673) $11x(1-4x)$

(674) $-11p(11p+11)$

(675) $12(3-2r)$

(676) $9(1-7a)$

(677) $x(1+6x)$

(678) $9b(8+7b)$

(679) $7n(n+11)$

(680) $5x(1-5x)$

Simplify each of the below expressions.

(681) $-10x - 12(4x - 9)$

(682) $-11n + 4(1 + 3n)$

(683) $-5m + 6(5m - 5)$

(684) $-6(x + 1) - 8$

(685) $8n + 11(12n + 10)$

(686) $-10(-6n - 2) + 5$

(687) $-8(5 - 7m) + 6$

(688) $-9(3 + 9n) + 2$

(689) $-3(1 - 4x) + 3$

(690) $-2 - 4(3 + 7x)$

BASIC ALGEBRA 1

Basic Math

Simplify each of the below expressions.

(691) $-6(-11 + a) + 4a$

(692) $-9 + 2(12m + 8)$

(693) $-3x - 4(x + 12)$

(694) $10 + 9(11m - 4)$

(695) $-6a + 10(10 + 8a)$

(696) $-12 + 8(5x - 6)$

(697) $-7 + 9(11n + 5)$

(698) $-3b + 7(b + 11)$

(699) $-5(12n + 8) + 12$

(700) $8n + 3(9n + 6)$

Simplify each of the below expressions.

(701) $-6x + 4(3 + x)$

(702) $-4(x + 4) - 4$

(703) $10(-2 - 6b) + 4$

(704) $-(6 - 5r) - 4$

(705) $11(x + 5) + 12x$

(706) $-5(1 - 15k) - 9(k + 18)$

(707) $15(-18r - 20) + 8(1 - 20r)$

(708) $15(1 - 2b) + 11(17 - 18b)$

(709) $6(7 + 19m) + 18(-12 + 8m)$

(710) $20(6 + 12x) - 2(1 - 6x)$

Simplify each of the below expressions.

(711) $7(9r - 2) + 9(1 - r)$

(712) $16(13 - 19x) + 7(14x - 15)$

(713) $20(p - 1) - 12(14 - 6p)$

(714) $7(9 - n) - 5(3n - 5)$

(715) $-16(12 + 9v) - 10(8v + 16)$

(716) $-6(7v - 20) - 17(16v - 1)$

(717) $-3(4v + 7) - 2(1 + 2v)$

(718) $12(1 - 16v) + 9(6v + 14)$

(719) $-13(b - 5) - 13(b + 3)$

(720) $20(13r - 12) - 7(7r - 1)$

BASIC ALGEBRA 1

Simplify each of the below expressions.

(721) $20(13v - 9) + 16(8v - 2)$

(722) $-8(n - 17) + 8(7n - 19)$

(723) $-2(9 + 19n) - 3(-5n - 11)$

(724) $5(1 - 13n) + 14(11 - 12n)$

(725) $16(3m + 10) + 7(9m - 11)$

(726) $-13(-9 + 16a) - (a - 7)$

(727) $15(x - 9) + 14(x + 16)$

(728) $-6(-12a + 11) - 5(1 + 19a)$

(729) $-6(1 - 17m) + 8(m + 9)$

(730) $-12(1 + 9x) - (1 - 6x)$

Solve each of the below equations.

(731) $b + 3 = -14$

(732) $k + 4 = 19$

(733) $-12 - n = -19$

(734) $n - 15 = -2$

(735) $17 - n = 31$

(736) $12 - n = 16$

(737) $8 - v = 23$

(738) $4 - m = -9$

(739) $m - 9 = -20$

(740) $b + 7 = -7$

Solve each of the below equations.

(741) $9 + x = -1$

(742) $4 + k = 11$

(743) $p - 17 = 0$

(744) $a + 1 = -5$

(745) $v - 8 = -20$

(746) $n - 19 = -36$

(747) $x - 8 = -22$

(748) $p + 8 = 9$

(749) $20 - n = 9$

(750) $n - 16 = -15$

Solve each of the below equations.

(751) $10 - r = 1$

(752) $n - 7 = -1$

(753) $3 - a = -10$

(754) $x + 7 = 20$

(755) $n + 15 = 0$

(756) $n + 32 = 2$

(757) $n - 8 = -1$

(758) $n + 28 = 2$

(759) $r + 4 = -24$

(760) $25 + m = 46$

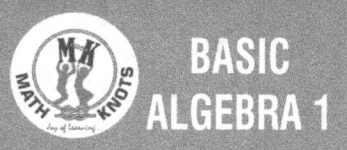

Solve each of the below equations.

(761) $v - 21 = -61$

(762) $x - 17 = -47$

(763) $r - 14 = 8$

(764) $r + 13 = 15$

(765) $x - 12 = -13$

(766) $k - 13 = -42$

(767) $x - 6 = 12$

(768) $x + 21 = -14$

(769) $-23 - n = -19$

(770) $-12 - x = 24$

Solve each of the below equations.

(771) $m + 16 = 32$

(772) $b - 1 = -4$

(773) $r - 20 = -43$

(774) $v - 13 = -37$

(775) $x + 13 = 23$

(776) $x + 3 = 14$

(777) $-23 - b = -9$

(778) $x + 34 = 27$

(779) $n + 21 = 45$

(780) $r - 12 = -16$

Solve each of the below equations.

(781) $8v = -96$

(782) $\dfrac{v}{12} = -\dfrac{3}{2}$

(783) $\dfrac{p}{20} = \dfrac{3}{10}$

(784) $-19k = -380$

(785) $\dfrac{m}{6} = 11$

(786) $18v = -180$

(787) $-16n = -64$

(788) $-17x = -119$

(789) $\dfrac{x}{18} = -11$

(790) $-18k = 162$

Solve each of the below equations.

(791) −13a = 247

(792) $\dfrac{n}{9} = \dfrac{7}{9}$

(793) −14n = 182

(794) 10k = 180

(795) $\dfrac{x}{10} = 2$

(796) −20n = 200

Solve each of the below equations.

(797) $\dfrac{x}{20} = 18$

(798) $\dfrac{x}{14} = -5$

(799) $4x = -60$

(800) $\dfrac{n}{13} = 7$

(801) $\dfrac{p}{15} = 16$

(802) $\dfrac{p}{17} = -\dfrac{11}{17}$

(803) $-11n = -198$

(804) $\dfrac{x}{8} = -2$

(805) $\dfrac{x}{15} = 6$

(806) $-12b = 24$

BASIC ALGEBRA 1

Basic Math

Solve each of the below equations.

(807) $16x = -176$

(808) $\dfrac{n}{4} = -\dfrac{19}{4}$

(809) $8n = 112$

(810) $\dfrac{p}{6} = -5$

(811) $\dfrac{a}{15} = -\dfrac{13}{15}$

(812) $2x = 28$

(813) $\dfrac{n}{11} = 9$

(814) $\dfrac{n}{10} = -\dfrac{19}{10}$

(815) $-17p = 34$

(816) $-6n = 42$

Solve each of the below equations.

(817) $\dfrac{x}{15} = \dfrac{6}{5}$

(818) $\dfrac{x}{18} = 5$

(819) $17n = -119$

(820) $-4a = -56$

(821) $\dfrac{a}{16} = 5$

(822) $-3v = 42$

(823) $\dfrac{r}{5} = \dfrac{7}{5}$

(824) $-18n = -270$

(825) $\dfrac{m}{12} = 15$

(826) $-17x = -34$

Solve each of the below equations.

(827) $\dfrac{p}{11} = -14$

(828) $11m = 11$

(829) $2m = 2$

(830) $-10x = -90$

(831) $5 - 2k = 37$

(832) $6 - 9v = -75$

(833) $\dfrac{p-3}{15} = -1$

(834) $1 + \dfrac{n}{3} = 6$

(835) $4 + 3n = -41$

(836) $\dfrac{x}{18} + 6 = 7$

Solve each of the below equations.

(837) $-7a - 6 = -111$

(838) $\dfrac{-1 + n}{2} = 4$

(839) $\dfrac{p + 4}{1} = 0$

(840) $-5 + \dfrac{v}{2} = -1$

(841) $\dfrac{-6 + b}{4} = -3$

(842) $1 + 2n = 25$

(843) $\dfrac{v}{12} + 8 = 7$

(844) $-6x + 3 = -3$

(845) $\dfrac{n}{9} + 10 = 9$

(846) $7k - 3 = -38$

Solve each of the below equations.

(847) $-4 + \dfrac{n}{9} = -3$

(848) $\dfrac{p-4}{2} = -9$

(849) $-3 + 9p = -57$

(850) $-10 - r = -19$

(851) $6 + 4k = -14$

(852) $-9 + \dfrac{x}{4} = -11$

(853) $\dfrac{r+6}{3} = 4$

(854) $\dfrac{m+3}{3} = -5$

(855) $-9 + 7k = 89$

(856) $5 + \dfrac{n}{4} = 6$

BASIC ALGEBRA 1

Basic Math

Solve each of the below equations.

(857) $8 - 8n = -80$

(858) $10m + 9 = -91$

(859) $\dfrac{-3 + b}{3} = 5$

(860) $-1 + \dfrac{n}{14} = 0$

(861) $1 - 2m = 33$

(862) $\dfrac{m - 2}{2} = -9$

(863) $\dfrac{n + 5}{10} = 2$

(864) $\dfrac{x}{3} - 6 = -3$

(865) $\dfrac{-3 + b}{21} = -1$

(866) $-1 + \dfrac{m}{10} = 0$

Solve each of the below equations.

(867) $\dfrac{3+x}{10} = 1$

(868) $-7n - 1 = 48$

(869) $2m - 5 = -33$

(870) $\dfrac{b+4}{7} = 3$

(871) $7 + \dfrac{r}{6} = 8$

(872) $\dfrac{p}{2} + 9 = 2$

(873) $\dfrac{a}{10} + 10 = 12$

(874) $\dfrac{-3+n}{2} = -2$

(875) $-10 + \dfrac{x}{8} = -9$

(876) $8p + 10 = 122$

Solve each of the below equations.

(877) $-2 - 2b = 34$

(878) $\dfrac{a+1}{8} = 2$

(879) $\dfrac{x+5}{3} = -3$

(880) $-10 + 10x = -150$

Solve each of the below inequalities and graph its solution.

(881) $x - 11 \leq 9$

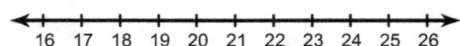

(882) $b + 7 \leq 4$

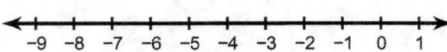

(883) $8 + x > -7$

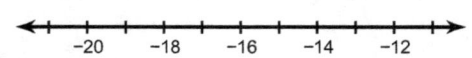

(884) $k - 19 \leq -31$

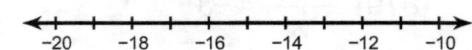

(885) $x - 13 \leq -13$

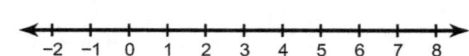

(886) $n - 14 \geq -2$

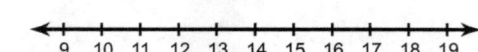

(887) $k + 16 > 33$

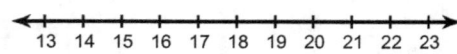

(888) $x - 15 \geq -24$

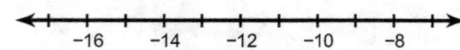

(889) $4 - x < -10$

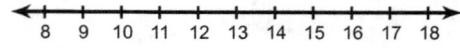

(890) $3 - x > 20$

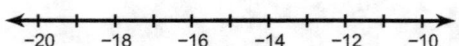

Solve each of the below inequalities and graph its solution.

(891) $11 - a > 29$

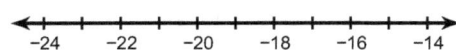

(892) $k + 11 \le -7$

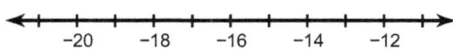

(893) $k + 20 \le 40$

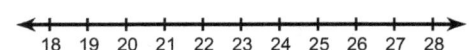

(894) $p - 9 \le 2$

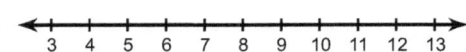

(895) $10 - m > 30$

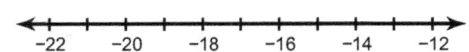

(896) $15 - v < 2$

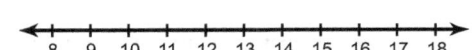

(897) $m - 4 \ge -10$

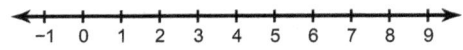

(898) $n - 19 < -20$

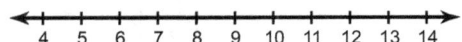

(899) $-13 + k > -7$

(900) $m + 2 \le 12$

Solve each of the below inequalities and graph its solution.

(901) $v + 17 \geq 22$

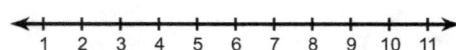

(902) $-7 - m \leq -8$

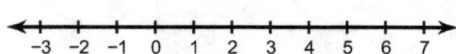

(903) $x - 17 \leq -19$

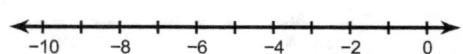

(904) $n - 4 \leq 9$

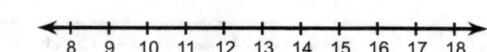

(905) $-11 + x > -10$

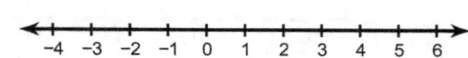

(906) $r + 2 > 18$

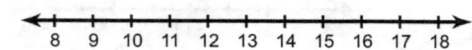

(907) $p - 8 < 2$

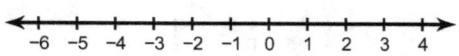

(908) $m - (-4) \geq -4$

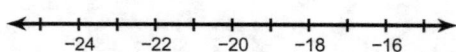

(909) $1 - b \leq 2$

(910) $-19 - r \leq 0$

Solve each of the below inequalities and graph its solution.

(911) $b + (-20) \geq -21$

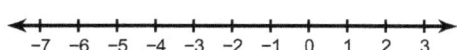

(912) $b + 2 > -13$

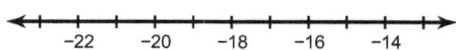

(913) $r - 12 > 7$

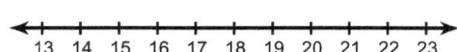

(914) $x + (-6) \leq -23$

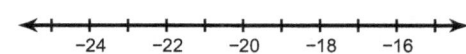

(915) $x + 14 \geq 7$

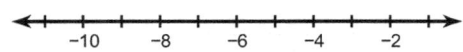

(916) $x - 18 \geq -7$

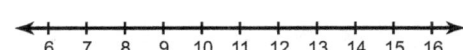

(917) $x - (-3) > 10$

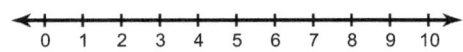

(918) $11 - m \leq -5$

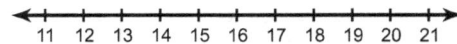

(919) $17 - n < 19$

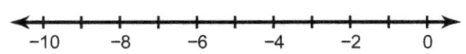

(920) $19 + r \leq 11$

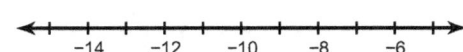

Solve each of the below inequalities and graph its solution.

(921) $18 + m \leq 37$

(922) $x - 14 < -1$

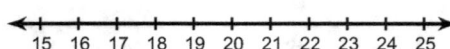

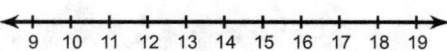

(923) $x - (-16) \geq -2$

(924) $n + (-19) \leq -6$

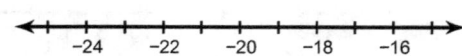

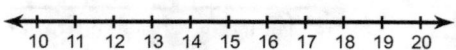

(925) $a + 7 > 18$

(926) $n + (-13) < -4$

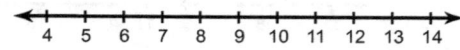

(927) $18 + v < 16$

(928) $x - (-16) \leq 13$

(929) $3 + b \geq -17$

(930) $b + 15 \geq -4$

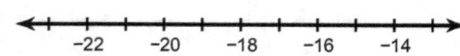

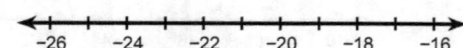

BASIC ALGEBRA 1

Solve each of the below inequalities and graph its solution.

(931) $-13p < 0$

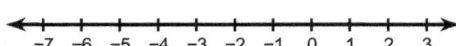

(932) $-132 \geq -11b$

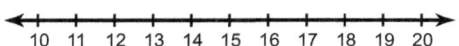

(933) $-\dfrac{5}{4} < \dfrac{x}{12}$

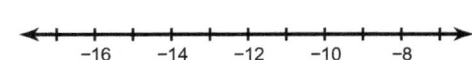

(934) $13n < -208$

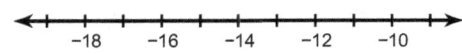

(935) $98 > -14b$

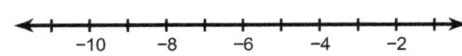

(936) $-15 < \dfrac{v}{8}$

(937) $18 \leq \dfrac{n}{3}$

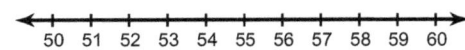

(938) $\dfrac{a}{19} \geq 13$

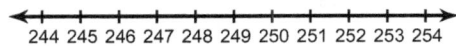

Solve each of the below inequalities and graph its solution.

(939) $9x \le 45$

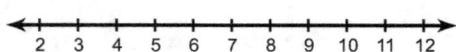

(940) $\dfrac{a}{20} \le 8$

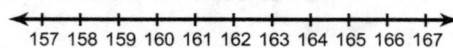

(941) $-136 < -8r$

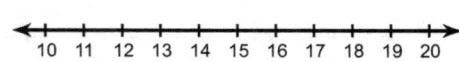

(942) $\dfrac{n}{6} < -14$

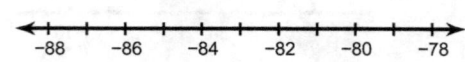

(943) $-15 \le \dfrac{r}{14}$

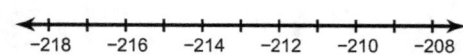

(944) $\dfrac{b}{6} < -19$

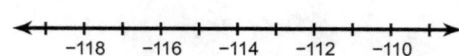

(945) $\dfrac{a}{9} \ge 15$

(946) $-12x \le 168$

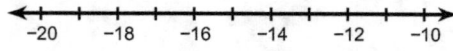

Solve each of the below inequalities and graph its solution.

(947) $13p > -78$

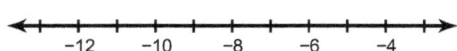

(948) $-90 > -6n$

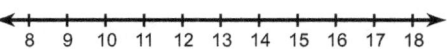

(949) $10 < \dfrac{n}{12}$

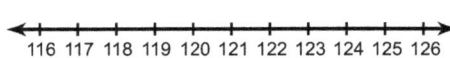

(950) $19x \leq 380$

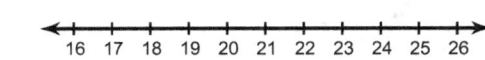

(951) $25 < -5m$

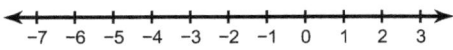

(952) $-90 < 6k$

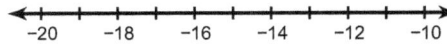

Solve each of the below inequalities and graph its solution.

(953) $-6 > \dfrac{k}{8}$

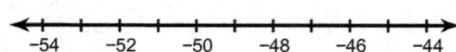

(954) $-19 \leq \dfrac{x}{18}$

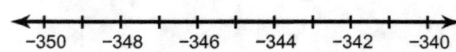

(955) $\dfrac{x}{6} \geq 17$

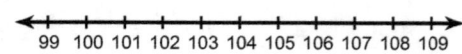

(956) $\dfrac{x}{20} > -5$

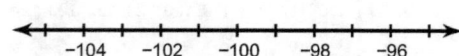

(957) $3m \geq -51$

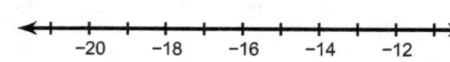

(958) $\dfrac{k}{15} > -15$

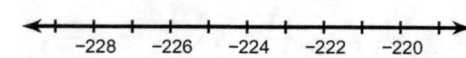

Solve each of the below inequalities and graph its solution.

(959) $18b \geq -162$

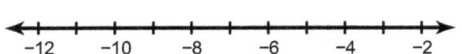

(960) $19v \leq 228$

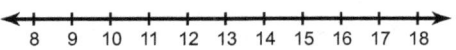

(961) $19 < \dfrac{v}{11}$

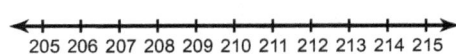

(962) $-6 \leq \dfrac{n}{4}$

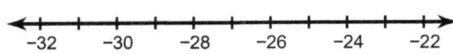

(963) $-160 < -20n$

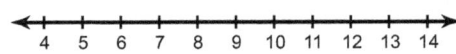

(964) $-160 \geq 8k$

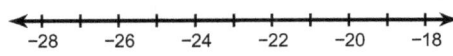

Solve each of the below inequalities and graph its solution.

(965) $-240 > -20x$

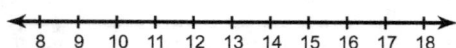

(966) $144 < 9x$

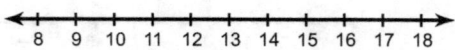

(967) $\dfrac{r}{7} > -\dfrac{12}{7}$

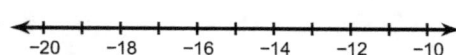

(968) $\dfrac{n}{4} \geq 10$

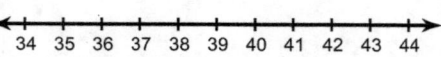

(969) $-27 < -9r$

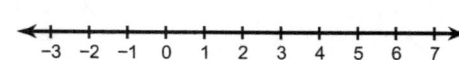

(970) $-20 \geq \dfrac{r}{11}$

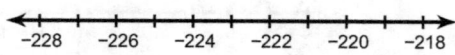

BASIC ALGEBRA 1

Solve each of the below inequalities and graph its solution.

(971) $-105 \geq 17x - 20$

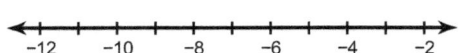

(972) $-12x - 17 \leq 115$

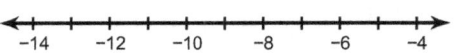

(973) $5 + 5x \leq 20$

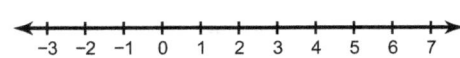

(974) $\dfrac{-10 + m}{39} \geq -1$

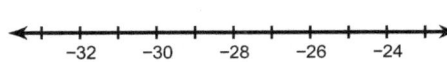

(975) $13p + 16 \geq -478$

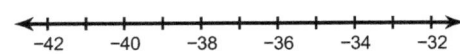

(976) $\dfrac{a - 5}{11} \geq -3$

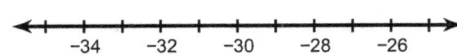

Solve each of the below inequalities and graph its solution.

(977) $-39 \geq 13 + 2k$

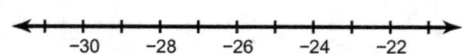

(978) $-4 + 12v \geq 32$

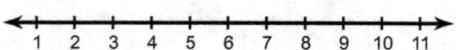

(979) $2 > -1 + \dfrac{n}{3}$

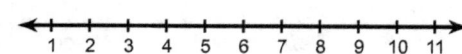

(980) $-7 \geq \dfrac{x + 16}{3}$

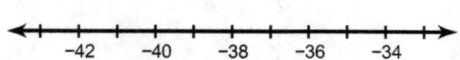

(981) $-11 > \dfrac{x}{8} - 15$

(982) $-2 < \dfrac{n - 2}{4}$

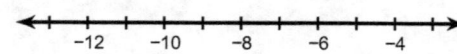

Solve each of the below inequalities and graph its solution.

(983) $-9r - 2 \leq 169$

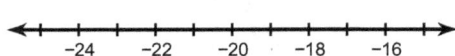

(984) $\dfrac{x + 10}{2} > 17$

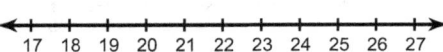

(985) $19 + \dfrac{x}{3} \geq 11$

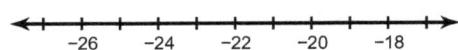

(986) $\dfrac{r}{17} + 14 < 16$

(987) $20 \leq 14 + \dfrac{x}{4}$

(988) $-225 \geq -18v + 9$

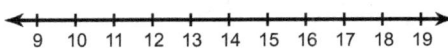

BASIC ALGEBRA 1

Basic Math

Solve each of the below inequalities and graph its solution.

(989) $2 < \dfrac{-3 + n}{2}$

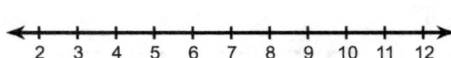

(990) $-13n - 7 < 58$

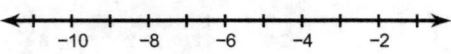

(991) $9 < 11 + \dfrac{p}{3}$

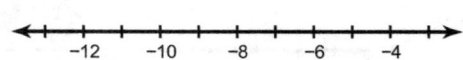

(992) $-13 \geq 17x - 13$

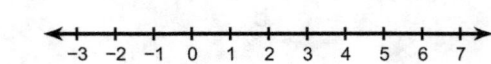

(993) $\dfrac{x}{4} - 2 < 2$

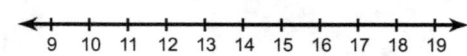

(994) $1 < \dfrac{18 + x}{45}$

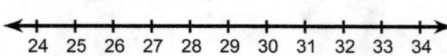

Solve each of the below inequalities and graph its solution.

(995) $-15 + \dfrac{n}{3} \le -7$

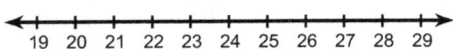

(996) $2 \le 3 + \dfrac{n}{34}$

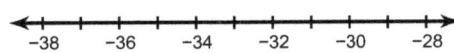

(997) $15 + \dfrac{n}{4} < 24$

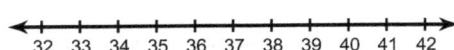

(998) $7 \le 9 + \dfrac{v}{9}$

(999) $-14v - 3 > -73$

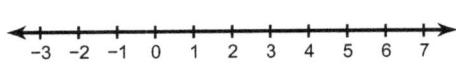

(1000) $-2 \ge \dfrac{k}{34} - 3$

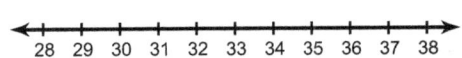

Solve each of the below inequalities and graph its solution.

(1001) $\dfrac{n+9}{3} > 9$

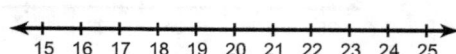

(1002) $-18 + 15b \le 492$

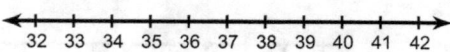

(1003) $11 + \dfrac{x}{2} \le -7$

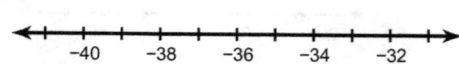

(1004) $18 > 16 + \dfrac{b}{5}$

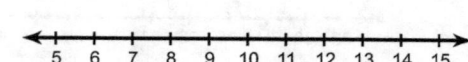

(1005) $\dfrac{r-2}{26} \ge -1$

(1006) $13 + \dfrac{v}{17} \ge 15$

Solve each of the below inequalities and graph its solution.

(1007) $\dfrac{a-5}{2} > -22$

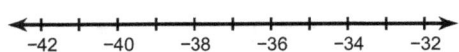

(1008) $-15 + \dfrac{x}{2} \geq -9$

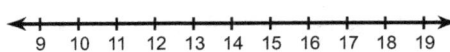

(1009) $\dfrac{2+x}{14} \leq 2$

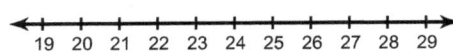

(1010) $\dfrac{-20+x}{2} \geq -9$

(1011) $-12 > -18 + \dfrac{b}{3}$

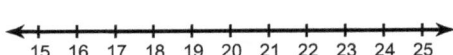

(1012) $197 > -11 + 8k$

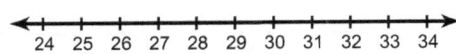

Solve each of the below inequalities and graph its solution.

(1013) $-20 + \dfrac{k}{10} \leq -21$

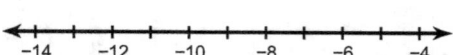

(1014) $\dfrac{x}{9} - 16 \leq -15$

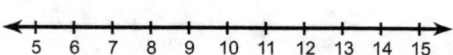

(1015) $6n - 8 > 196$

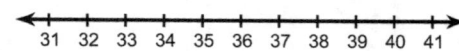

(1016) $4x + 11 > 135$

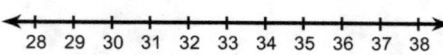

(1017) $14 + 5x \geq 89$

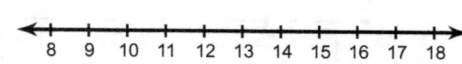

(1018) $\dfrac{p - 19}{29} < -2$

Solve each of the below inequalities and graph its solution.

(1019) $-17 \geq -12 + \dfrac{v}{5}$

(1020) $\dfrac{x}{4} + 13 \leq 14$

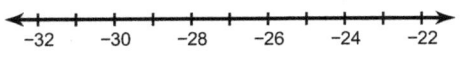

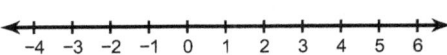

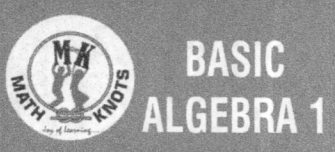

BASIC ALGEBRA 1

Answer Key

(1) $n + 12$ (2) $7 + n$ (3) $n + 9$ (4) $n + 5$

(5) $12 + n$ (6) $n + 11$ (7) $9 + n$ (8) $n + 10$

(9) $2 + n$ (10) $12 + 6$ (11) $n + 12$ (12) $3 + n$

(13) $n + 12$ (14) $n + 8$ (15) $6 + 6$ (16) $11 + 12$

(17) $7 + 11$ (18) $n + 11$ (19) $8 + n$ (20) $n + 6$

(21) $11 - 9$ (22) $25 - 21$ (23) $n - 15$ (24) $w - 17$

(25) $n - 9$ (26) $23 - n$ (27) $12 - 3$ (28) $20 - r$

BASIC ALGEBRA 1

Basic Math Answer Keys

(29) $15 - n$ (30) $n - 15$ (31) $n - 6$ (32) $20 - 19$

(33) $13 - 9$ (34) $27 - 14$ (35) $p - 3$ (36) $n - 7$

(37) $n - 25$ (38) $13 - 7$ (39) $12 - 10$ (40) $12 - 10$

(41) $n \cdot 9$ (42) $2n$ (43) $8 \cdot 5$ (44) $n \cdot 7$

(45) $6n$ (46) $2 \cdot 11$ (47) $n \cdot 6$ (48) $n \cdot 11$

(49) $11n$ (50) $2n$ (51) $8 \cdot 8$ (52) $2 \cdot 12$

(53) $5 \cdot 10$ (54) $2 \cdot 5$ (55) $8n$ (56) $8n$

(57) $2 \cdot 10$ (58) $2 \cdot 10$ (59) $2n$ (60) $3 \cdot 8$

(61) $\dfrac{16}{2}$ (62) $\dfrac{24}{4}$ (63) $\dfrac{n}{2}$ (64) $\dfrac{n}{4}$

(65) $\dfrac{6}{n}$ (66) $\dfrac{14}{2}$ (67) $\dfrac{14}{2}$ (68) $\dfrac{33}{3}$

(69) $\dfrac{6}{2}$ (70) $\dfrac{n}{4}$ (71) $\dfrac{8}{2}$ (72) $\dfrac{n}{3}$

(73) $\dfrac{36}{4}$ (74) $\dfrac{6}{2}$ (75) $\dfrac{33}{n}$ (76) $\dfrac{20}{4}$

(77) $\dfrac{25}{5}$ (78) $\dfrac{n}{2}$ (79) $\dfrac{n}{8}$ (80) $\dfrac{16}{2}$

(81) 5^2 (82) 3^2 (83) 9^2 (84) p^2

BASIC ALGEBRA 1

(85) 3^4 (86) 8^2 (87) 6^2 (88) u^3

(89) 2^2 (90) 5^3 (91) 10^y (92) 4^3

(93) 14^y (94) 11^n (95) 3^y (96) 9^n

(97) 4^3 (98) 3^3 (99) 5^3 (100) n^3

(101) $n + 8 = 45$ (102) $n + 8 = 39$ (103) $n + 10 = 24$ (104) $n + 9 = 7$

(105) $n + 11 = 34$ (106) $n + 11 = 14$ (107) $n + 6 = 27$ (108) $n + 10 = 48$

(109) $n + 12 = 16$ (110) $n + 10 = 19$ (111) $n + 9 = 33$ (112) $n + 5 = 24$

BASIC ALGEBRA 1

Basic Math Answer Keys

(113) $n + 8 = 12$ (114) $n + 5 = 29$ (115) $n + 9 = 24$ (116) $n + 7 = 18$

(117) $n + 7 = 31$ (118) $n + 5 = 17$ (119) $n + 9 = 30$ (120) $n + 7 = 18$

(121) $n - 12 = 8$ (122) $d - 9 = 22$ (123) $n - 10 = 6$ (124) $n - 10 = 30$

(125) $n - 14 = 24$ (126) $n - 13 = 20$ (127) $n - 6 = 13$ (128) $n - 4 = 40$

(129) $n - 9 = 36$ (130) $n - 8 = 48$ (131) $n - 14 = 6$ (132) $n - 16 = 48$

(133) $k - 3 = 33$ (134) $m - 8 = 47$ (135) $n - 7 = 32$ (136) $n - 4 = 50$

(137) $x - 4 = 18$ (138) $n - 16 = 27$ (139) $n - 8 = 31$ (140) $w - 6 = 19$

BASIC ALGEBRA 1

(141) $n \cdot 12 = 48$ (142) $n \cdot 9 = 18$ (143) $n \cdot 7 = 37$ (144) $n \cdot 5 = 42$

(145) $n \cdot 8 = 44$ (146) $n \cdot 10 = 9$ (147) $n \cdot 6 = 9$ (148) $n \cdot 5 = 30$

(149) $2n = 6$ (150) $n \cdot 6 = 43$ (151) $n \cdot 7 = 19$ (152) $n \cdot 11 = 15$

(153) $n \cdot 10 = 46$ (154) $n \cdot 8 = 21$ (155) $n \cdot 12 = 40$ (156) $\dfrac{n}{3} = 46$

(157) $\dfrac{n}{5} = 30$ (158) $\dfrac{n}{4} = 32$ (159) $\dfrac{n}{2} = 16$ (160) $\dfrac{n}{7} = 20$

(161) $\dfrac{n}{6} = 20$ (162) $\dfrac{n}{2} = 47$ (163) $\dfrac{n}{7} = 32$ (164) $\dfrac{n}{6} = 16$

(165) $\dfrac{n}{4} = 44$ (166) $\dfrac{n}{8} = 44$ (167) $\dfrac{n}{2} = 47$ (168) $\dfrac{n}{8} = 36$

BASIC ALGEBRA 1

(169) $\dfrac{n}{5} = 36$ (170) $\dfrac{n}{3} = 42$ (171) $q^4 = 40$ (172) $4^v = 44$

(173) $7^n = 33$ (174) $x^6 = 28$ (175) $5^m = 15$ (176) $t^3 = 35$

(177) $a^5 = 47$ (178) $t^9 = 20$ (179) $x^7 = 22$ (180) $15^n = 50$

(181) $9^n = 30$ (182) $n^3 = 42$ (183) $n^9 = 10$ (184) $n^2 = 27$

(185) $x^8 = 20$ (186) $12^n = 30$ (187) $n^6 = 50$ (188) $n^7 = 42$

(189) $5^k = 46$ (190) $y^4 = 46$ (191) $n + 7 \geq 23$ (192) $n - 10 \geq 13$

(193) $n + 12 \leq 45$ (194) $n - 5 < 28$ (195) $n + 12 \leq 50$ (196) $n - 16 \geq 46$

BASIC ALGEBRA 1

(197) $n - 9 < 22$

(198) $n + 12 > 48$

(199) $x - 17 \leq 15$

(200) $n - 9 > 11$

(201) $n + 10 \leq 19$

(202) $n + 8 \leq 13$

(203) $n - 20 < 34$

(204) $n - 9 \geq 15$

(205) $n - 11 \geq 24$

(206) $n - 15 < 18$

(207) $n + 10 > 20$

(208) $n + 5 \geq 36$

(209) $n - 6 > 35$

(210) $n + 5 < 12$

(211) $n + 11 \leq 33$

(212) $n - 9 \leq 31$

(213) $n + 9 \leq 21$

(214) $n + 8 \leq 50$

(215) $n - 23 < 32$

(216) $n - 20 \geq 38$

(217) $n - 8 < 16$

(218) $n + 11 \geq 8$

(219) $n + 9 < 10$

(220) $w - 3 > 12$

(221) $n + 9 < 46$

(222) $n + 12 \geq 20$

(223) $n - 15 \geq 27$

(224) $a - 11 \leq 15$

BASIC ALGEBRA 1

(225) $n + 5 \leq 18$ (226) $n + 5 < 21$ (227) $n + 10 \leq 14$ (228) $n + 8 \geq 13$

(229) $n - 12 \leq 29$ (230) $n + 11 > 26$ (231) $\dfrac{n}{8} \leq 10$ (232) $\dfrac{n}{6} \geq 14$

(233) $2n \geq 29$ (234) $n^3 > 6$ (235) $n \cdot 12 \geq 24$ (236) $n \cdot 8 > 24$

(237) $\dfrac{n}{7} \geq 17$ (238) $n \cdot 7 \leq 43$ (239) $n^2 \leq 50$ (240) $\dfrac{n}{2} \geq 31$

(241) $\dfrac{n}{2} \leq 11$ (242) $n \cdot 6 \leq 43$ (243) $\dfrac{n}{4} > 39$ (244) $n \cdot 6 < 16$

(245) $n \cdot 11 < 16$ (246) $2n < 10$ (247) $2n \leq 13$ (248) $n \cdot 9 \leq 47$

(249) $n \cdot 7 \geq 12$ (250) $n \cdot 7 \geq 26$ (251) $\dfrac{n}{2} > 43$ (252) $\dfrac{n}{2} < 40$

(253) $\dfrac{n}{3} < 22$

(254) $n^2 < 22$

(255) $\dfrac{n}{3} > 45$

(256) $n^2 > 17$

(257) $\dfrac{n}{8} < 48$

(258) $n \cdot 12 > 20$

(259) $\dfrac{n}{7} \le 10$

(260) $n^3 < 48$

(261) $\dfrac{n}{3} < 30$

(262) $\dfrac{n}{4} < 40$

(263) $\dfrac{n}{4} > 10$

(264) $\dfrac{n}{2} \ge 17$

(265) $\dfrac{n}{5} \le 40$

(266) $\dfrac{n}{6} \le 41$

(267) $\dfrac{n}{6} > 39$

(268) $n \cdot 6 \ge 19$

(269) $\dfrac{n}{4} \le 39$

(270) $\dfrac{n}{5} \ge 18$

(271) $n^2 \le 17$

(272) $9^p < 49$

(273) $n^3 < 41$

(274) $p^9 < 45$

(275) $n^3 \le 28$

(276) $n^3 > 47$

(277) $v^{10} > 9$

(278) $10^x \le 36$

(279) $n^9 > 21$

(280) $n^3 \ge 49$

BASIC ALGEBRA 1

(281) $5^n < 22$ (282) $n^2 \geq 15$ (283) $x^8 \geq 39$ (284) $n^2 < 39$

(285) $n^6 \geq 37$ (286) $11^c > 49$ (287) $6^n \geq 27$ (288) $15^t \leq 7$

(289) $n^2 \geq 33$ (290) $15^w < 12$

(291) The difference of 30 and 19

(292) A number plus 10

293) 8 plus 6

(294) The difference of 16 and 4

(295) The sum of a number and 5

(296) The sum of 11 and a number

(297) 12 plus 9

(298) 11 increased by a number

(299) 11 less than n

BASIC ALGEBRA 1

(300) 8 increased by 11

(301) 7 less than 12

(302) 5 plus a number

(303) The difference of 23 and 4

(304) The difference of a number and 8

(305) 4 plus a number

(306) The sum of a number and 7

(307) 4 increased by a number

(308) 6 more than 11

(309) 8 plus a number

(310) A number decreased by 7

(311) The sum of a number and 11

(312) The sum of 8 and a number

(313) 18 decreased by 11

BASIC ALGEBRA 1

Basic Math Answer Keys

(314) 7 more than a number

(315) 9 plus 9

(316) 15 less than 21

(317) 8 increased by a number

(318) 6 less than c

(319) The difference of 10 and 5

(320) 12 less than 29

(321) A number plus 9

(322) r less than 12

(323) 30 minus 21

(324) A number minus 6

(325) 4 less than 15

(326) 7 more than 11

(327) A number plus 8

(328) 8 more than a number

(329) 12 more than 10

(330) A number minus 12

(331) The product of 4 and 6

(332) 40 divided by 8

(333) Twice 6

(334) The product of 4 and a number

(335) Half of a number

(336) The product of a number and 10

(337) 11 times 6

(338) The quotient of a number and 6

(339) The quotient of 77 and 7

(340) A number divided by 3

(341) Half of 16

BASIC ALGEBRA 1

(342) Twice 12

(343) Half of 20

(344) A number times 8

(345) Half of 14

(346) The quotient of 42 and 6

(347) Half of 12

(348) Twice a number

(349) The quotient of 16 and 8

(350) The quotient of a number and 4

(351) 4 times a number

(352) A number divided by 5

(353) 36 divided by a number

(354) A number times 6

(355) Twice 3

BASIC ALGEBRA 1

(356) 4 divided by a number

(357) The product of 2 and 7

(358) Twice 2

(359) Half of 10

(360) Twice 9

(361) 77 divided by 7

(362) A number divided by 6

(363) The quotient of a number and 2

(364) Twice 10

(365) 3 times a number

(366) 12 divided by a number

(367) 8 times 9

BASIC ALGEBRA 1

(368) Half of 8

(369) The product of 6 and 8

(370) A number times 5

(371) Twice 8

(372) The product of 2 and 12

(373) 12 times a number

(374) 11 times 9

(375) The quotient of 4 and 2

(376) The product of a number and 11

(377) The quotient of 30 and a number

(378) The quotient of 60 and a number

(379) 2 times a number

(380) Twice 5

(381) A number cubed

(382) A number squared

(383) 6 to the 2nd

(384) The 8th power of n

(385) 10 squared

(386) The 2nd power of 10

(387) 9 to the tth power

(388) 5 cubed

(389) 3 cubed

(390) 5 squared

(391) The c power of 4

(392) 5 cubed

(393) 7 squared

(394) 4 squared

(395) 5 squared

BASIC ALGEBRA 1

(396) 6 squared

(397) 2 cubed

(398) 12 to the nth power

(399) The 6th power of d

(400) 3 cubed

(401) A number decreased by 9 is equal to 44

(402) The sum of a number and 7 is equal to 27

(403) The difference of a number and 12 is 14

(404) 21 less than u is equal to 46

(405) A number increased by 11 is equal to 45

BASIC ALGEBRA 1

(406) The difference of a number and 6 is 34

(407) 8 less than n is 47

(408) A number times 11 is 37

(409) A number times 7 is 31

(410) A number decreased by 6 is 50

(411) A number decreased by 20 is equal to 46

(412) The difference of a number and 5 is equal to 18

(413) A number times 6 is 43

(414) The difference of a number and 13 is equal to 40

(415) A number increased by 6 is equal to 13

(416) A number increased by 5 is equal to 48

(417) The product of a number and 9 is equal to 34

(418) 3 less than n is 11

(419) A number minus 9 is equal to 22

(420) A number decreased by 22 is 29

(421) The sum of a number and 11 is 24

(422) 7 more than a number is 36

(423) A number increased by 12 is equal to 34

(424) Twice a number is 19

(425) A number minus 12 is equal to 18

(426) The difference of a number and 3 is equal to 46

(427) A number times 5 is 44

(428) A number decreased by 12 is equal to 45

(429) A number increased by 9 is equal to 30

(430) A number decreased by 17 is 33

(431) A number minus 4 is equal to 37

(432) The product of a number and 5 is 17

(433) A number increased by 8 is equal to 30

(434) 9 more than a number is 20

(435) 12 more than a number is 17

(436) The product of a number and 7 is equal to 22

(437) A number minus 3 is equal to 29

(438) The sum of a number and 5 is equal to 43

(439) The product of a number and 12 is 20

(440) A number minus 22 is 34

(441) 14 less than m is 8

(442) A number minus 14 is equal to 8

(443) 11 more than a number is equal to 16

(444) The difference of a number and 7 is 49

(445) The difference of a number and 20 is 8

(446) The sum of a number and 12 is equal to 15

(447) The sum of a number and 9 is 33

(448) 12 less than d is 12

(449) 4 less than u is equal to 13

(450) The sum of a number and 8 is equal to 5

(451) w to the 10th is 40

(452) A number divided by 8 is 15

(453) The m power of 12 is 42

(454) 13 to the n is equal to 40

(455) 8 to the x is equal to 34

(456) 4 to the x is 15

(457) A number cubed is equal to 49

(458) The quotient of a number and 7 is equal to 33

(459) Half of n is equal to 25

(460) n squared is 46

(461) The quotient of n and 5 is equal to 27

(462) The 8th power of w is equal to 20

(463) The quotient of x and 6 is 34

(464) p to the 2nd is equal to 7

(465) The quotient of x and 2 is equal to 6

(466) 14 to the x is equal to 46

(467) The u power of 6 is equal to 27

(468) The 3rd power of r is equal to 15

(469) The m power of 14 is 41

(470) The quotient of x and 4 is 36

(471) n times 7 is greater than or equal to 36

(472) The quotient of a number and 3 is less than or equal to 47

(473) r minus 6 is less than or equal to 44

(474) The quotient of x and 8 is greater than 23

(475) 8 more than a number is less than 6

(476) The quotient of a number and 6 is less than or equal to 44

(477) A number decreased by 24 is greater than 12

(478) r squared is less than or equal to 26

(479) 10 more than m is less than or equal to 29

(480) r to the 5th is greater than or equal to 40

(481) z divided by 6 is less than or equal to 13

(482) 5 more than n is greater than 39

(483) u cubed is less than 32

(484) x divided by 3 is less than 11

(485) Twice n is greater than 28

(486) The quotient of u and 8 is less than or equal to 22

(487) b increased by 9 is greater than 20

(488) n minus 16 is less than 32

(489) The sum of n and 8 is less than or equal to 49

(490) 19 less than y is greater than or equal to 15

(491) A number minus 10 is greater than or equal to 45

(492) 7 more than d is greater than or equal to 8

(493) The 8th power of c is greater than or equal to 49

(494) Twice b is less than 14

(495) z divided by 7 is greater than or equal to 44

(496) 6 less than z is greater than 24

(497) n times 8 is greater than 5

(498) A number squared is less than 23

(499) The sum of a number and 12 is less than or equal to 14

(500) The quotient of a number and 2 is less than or equal to 33

(501) A number squared is greater than 40

(502) The quotient of a number and 7 is greater than or equal to 40

(503) A number minus 18 is less than or equal to 27

(504) n to the 6th is greater than 12

(505) The product of a number and 12 is less than 8

(506) The difference of n and 9 is greater than or equal to 30

(507) The sum of n and 9 is greater than 5

(508) A number times 5 is less than or equal to 48

(509) c minus 9 is greater than or equal to 13

(510) Half of n is less than 10

(511) The product of y and 6 is less than 32

(512) The r power of 6 is less than 7

(513) 5 less than q is less than or equal to 22

(514) A number times 12 is greater than 24

(515) 20 less than p is greater than 13

(516) 10 more than a number is greater than or equal to 21

(517) 9 to the x is less than 37

(518) The product of a number and 5 is less than or equal to 19

(519) The quotient of c and 7 is greater than 30

BASIC ALGEBRA 1

(520) y decreased by 8 is greater than 6

(521) Twice n is greater than or equal to 7

(522) The product of x and 11 is less than or equal to 20

(523) The product of x and 10 is less than 49

(524) The u power of 7 is greater than 5

(525) 6 more than a number is greater than or equal to 38

(526) The product of b and 11 is greater than or equal to 37

(527) u divided by 5 is less than 11

(528) m cubed is greater than 10

(529) y decreased by 7 is less than 37

(530) r decreased by 9 is greater than or equal to 42

BASIC ALGEBRA 1

Basic Math Answer Keys

(531) 4   (532) 3   (533) 6   (534) 7

(535) 4   (536) 5   (537) 5   (538) 1

(539) 3   (540) 8   (541) 10   (542) 44

(543) 6   (544) 1   (545) 25   (546) 10

(547) 52   (548) 12   (549) 10   (550) 1

(551) 10   (552) 1   (553) 20   (554) 46

(555) 3   (556) 9   (557) 11   (558) 3

BASIC ALGEBRA 1

Basic Math Answer Keys

(559) 3   (560) 17   (561) 2   (562) 50

(563) 58   (564) 3   (565) 2   (566) 2

(567) 112   (568) 10   (569) 13   (570) 14

(571) 33   (572) 88   (573) 46   (574) 15

(575) 66   (576) 10   (577) 120   (578) 2

(579) 10   (580) 79   (581) 6   (582) 1

(583) 64   (584) 50   (585) 35   (586) 15

BASIC ALGEBRA 1

Basic Math Answer Keys

(587) 9   (588) 8   (589) 8   (590) 4

(591) 9   (592) 1   (593) 14   (594) 9

(595) 9   (596) 100   (597) 3   (598) 2

(599) 19   (600) 20   (601) 4   (602) 11

(603) 2   (604) 1   (605) 42   (606) 1

(607) 16   (608) 64   (609) 10   (610) 78

(611) 22   (612) 5   (613) 92   (614) 14

BASIC ALGEBRA 1

Basic Math Answer Keys

(615) 31 (616) 3 (617) 4 (618) 80

(619) 18 (620) 67 (621) 18 (622) 10

(623) 23 (624) 112 (625) 11 (626) 19

(627) 80 (628) 22 (629) 5 (630) 5

(631) n (632) $5a + 8$ (633) $-3x - 1$ (634) $-n + 2$

(635) $2x$ (636) $11 - r$ (637) $4r$ (638) $6b$

(639) $-1 + 8x$ (640) 2 (641) $7p$ (642) $-12x$

BASIC ALGEBRA 1

Basic Math Answer Keys

(643) −14n   (644) −3n − 4   (645) −1 + 9n   (646) 8x

(647) 3 + 2x   (648) 17x   (649) −10n   (650) −7x

(651) 4x   (652) 9b   (653) 2b − 8   (654) −5n

(655) −17x   (656) $22k^2 + 88k$ (657) −3 − 27m   (658) $-2m^2 + 2m$

(659) $-3b^2 + 30b$ (660) 70 − 28v   (661) 63x − 99   (662) −84 + 77k

(663) −20p + 25   (664) $-30x^2 - 6x$ (665) 44v + 12   (666) $-30k^2 - 66k$

(667) $-9a - 81a^2$ (668) 108x + 84   (669) −20n − 30   (670) $-5p^2 + 5p$

BASIC ALGEBRA 1

Basic Math Answer Keys

(671) $-11 + 99k$ (672) $-10r - 6$ (673) $11x - 44x^2$ (674) $-121p^2 - 121p$

(675) $36 - 24r$ (676) $9 - 63a$ (677) $x + 6x^2$ (678) $72b + 63b^2$

(679) $7n^2 + 77n$ (680) $5x - 25x^2$ (681) $-58x + 108$ (682) $n + 4$

(683) $25m - 30$ (684) $-6x - 14$ (685) $140n + 110$ (686) $60n + 25$

(687) $-34 + 56m$ (688) $-25 - 81n$ (689) $12x$ (690) $-14 - 28x$

(691) $66 - 2a$ (692) $7 + 24m$ (693) $-7x - 48$ (694) $-26 + 99m$

(695) $74a + 100$ (696) $-60 + 40x$ (697) $38 + 99n$ (698) $4b + 77$

BASIC ALGEBRA 1

Basic Math Answer Keys

(699) −60n − 28   (700) 35n + 18   (701) −2x + 12   (702) −4x − 20

(703) −16 − 60b   (704) −10 + 5r   (705) 23x + 55   (706) −167 + 66k

(707) −430r − 292   (708) 202 − 228b   (709) −174 + 258m   (710) 118 + 252x

(711) 54r − 5   (712) 103 − 206x   (713) 92p − 188   (714) 88 − 22n

(715) −352 − 224v   (716) −314v + 137   (717) −16v − 23   (718) 138 − 138v

(719) −26b + 26   (720) 211r − 233   (721) 388v − 212   (722) 48n − 16

(723) 15 − 23n   (724) 159 − 233n   (725) 111m + 83   (726) 124 − 209a

BASIC ALGEBRA 1

Basic Math Answer Keys

(727) $29x + 89$ (728) $-23a - 71$ (729) $66 + 110m$ (730) $-13 - 102x$

(731) -17 (732) 15 (733) 7 (734) 13

(735) -14 (736) -4 (737) -15 (738) 13

(739) -11 (740) -14 (741) -10 (742) 7

(743) 17 (744) -6 (745) -12 (746) -17

(747) -14 (748) 1 (749) 11 (750) 1

(751) 9 (752) 6 (753) 13 (754) 13

BASIC ALGEBRA 1

Basic Math Answer Keys

(755) −15   (756) −30   (757) 7   (758) −26

(759) −28   (760) 21   (761) −40   (762) −30

(763) 22   (764) 2   (765) −1   (766) −29

(767) 18   (768) −35   (769) −4   (770) −36

(771) 16   (772) −3   (773) −23   (774) −24

(775) 10   (776) 11   (777) −14   (778) −7

(779) 24   (780) −4   (781) −12   (782) −18

BASIC ALGEBRA 1

Basic Math Answer Keys

(783) 6   (784) 20   (785) 66   (786) −10

(787) 4   (788) 7   (789) −198   (790) −9

(791) −19   (792) 7   (793) −13   (794) 18

(795) 20   (796) −10   (797) 360   (798) −70

(799) −15   (800) 91   (801) 240   (802) −11

(803) 18   (804) −16   (805) 90   (806) −2

(807) −11   (808) −19   (809) 14   (810) −30

BASIC ALGEBRA 1

Basic Math Answer Keys

(811) −13   (812) 14   (813) 99   (814) −19

(815) −2   (816) −7   (817) 18   (818) 90

(819) −7   (820) 14   (821) 80   (822) −14

(823) 7   (824) 15   (825) 180   (826) 2

(827) −154   (828) 1   (829) 1   (830) 9

(831) −16   (832) 9   (833) −12   (834) 15

(835) −15   (836) 18   (837) 15   (838) 9

BASIC ALGEBRA 1

Basic Math Answer Keys

(839) −4   (840) 8   (841) −6   (842) 12

(843) −12   (844) 1   (845) −9   (846) −5

(847) 9   (848) −14   (849) −6   (850) 9

(851) −5   (852) −8   (853) 6   (854) −18

(855) 14   (856) 4   (857) 11   (858) −10

(859) 18   (860) 14   (861) −16   (862) −16

(863) 15   (864) 9   (865) −18   (866) 10

BASIC ALGEBRA 1

Basic Math Answer Keys

(867) 7   (868) −7   (869) −14   (870) 17

(871) 6   (872) −14   (873) 20   (874) −1

(875) 8   (876) 14   (877) −18   (878) 15

(879) −14   (880) −14   (881) $x \leq 20$:

(882) $b \leq -3$:   (883) $x > -15$: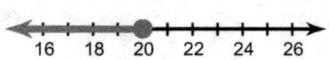

(884) $k \leq -12$:   (885) $x \leq 0$:

(886) $n \geq 12$: 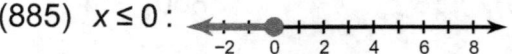   (887) $k > 17$: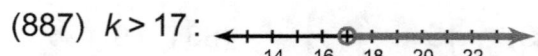

Basic Math Answer Keys

(888) $x \geq -9$:

(889) $x > 14$:

(890) $x < -17$:

(891) $a < -18$:

(892) $k \leq -18$:

(893) $k \leq 20$:

(894) $p \leq 11$:

(895) $m < -20$:

(896) $v > 13$:

(897) $m \geq -6$:

(898) $n < -1$:

(899) $k > 6$:

(900) $m \leq 10$:

(901) $v \geq 5$:

BASIC ALGEBRA 1

Basic Math Answer Keys

(902) $m \geq 1$:

(903) $x \leq -2$:

(904) $n \leq 13$:

(905) $x > 1$:

(906) $r > 16$:

(907) $p < 10$:

(908) $m \geq -8$:

(909) $b \geq -1$:

(910) $r \geq -19$:

(911) $b \geq -1$:

(912) $b > -15$:

(913) $r > 19$:

(914) $x \leq -17$:

(915) $x \geq -7$:

BASIC ALGEBRA 1

Basic Math Answer Keys

(916) $x \geq 11$:

(917) $x > 7$:

(918) $m \geq 16$:

(919) $n > -2$:

(920) $r \leq -8$:

(921) $m \leq 19$:

(922) $x < 13$:

(923) $x \geq -18$:

(924) $n \leq 13$:

(925) $a > 11$:

(926) $n < 9$:

(927) $v < -2$:

(928) $x \leq -3$:

(929) $b \geq -20$:

©All rights reserved-Math-Knots LLC., VA-USA
For more practice visit www.a4ace.com

www.math-knots.com

BASIC ALGEBRA 1

Basic Math Answer Keys

(930) $b \geq -19$:

(931) $p > 0$:

(932) $b \geq 12$:

(933) $x > -15$:

(934) $n < -16$:

(935) $b > -7$:

(936) $v > -120$:

(937) $n \geq 54$:

(938) $a \geq 247$:

(939) $x \leq 5$:

(940) $a \leq 160$:

(941) $r < 17$:

(942) $n < -84$:

(943) $r \geq -210$:

BASIC ALGEBRA 1

Basic Math Answer Keys

(944) $b < -114$:

(945) $a \geq 135$:

(946) $x \geq -14$:

(947) $p > -6$:

(948) $n > 15$:

(949) $n > 120$:

(950) $x \leq 20$:

(951) $m < -5$:

(952) $k > -15$:

(953) $k < -48$:

(954) $x \geq -342$:

(955) $x \geq 102$:

(956) $x > -100$:

(957) $m \geq -17$:

BASIC ALGEBRA 1

Basic Math Answer Keys

(958) $k > -225$:

(959) $b \geq -9$:

(960) $v \leq 12$:

(961) $v > 209$:

(962) $n \geq -24$:

(963) $n < 8$:

(964) $k \leq -20$:

(965) $x > 12$:

(966) $x > 16$:

(967) $r > -12$:

(968) $n \geq 40$:

(969) $r < 3$:

(970) $r \leq -220$:

(971) $x \leq -5$:

(972) $x \geq -11$:

(973) $x \leq 3$:

(974) $m \geq -29$:

(975) $p \geq -38$:

(976) $a \geq -28$:

(977) $k \leq -26$:

(978) $v \geq 3$:

(979) $n < 9$:

(980) $x \leq -37$:

(981) $x < 32$:

(982) $n > -6$:

(983) $r \geq -19$:

(984) $x > 24$:

(985) $x \geq -24$:

BASIC ALGEBRA 1

Basic Math Answer Keys

(986) $r < 34$:

(987) $x \geq 24$:

(988) $v \geq 13$:

(989) $n > 7$:

(990) $n > -5$:

(991) $p > -6$:

(992) $x \leq 0$:

(993) $x < 16$:

(994) $x > 27$:

(995) $n \leq 24$:

(996) $n \geq -34$:

(997) $n < 36$:

(998) $v \geq -18$:

(999) $v < 5$:

BASIC ALGEBRA 1

Basic Math Answer Keys

(1000) $k \leq 34$:

(1001) $n > 18$:

(1002) $b \leq 34$:

(1003) $x \leq -36$:

(1004) $b < 10$:

(1005) $r \geq -24$:

(1006) $v \geq 34$:

(1007) $a > -39$:

(1008) $x \geq 12$:

(1009) $x \leq 26$:

(1010) $x \geq 2$:

(1011) $b < 18$:

(1012) $k < 26$:

(1013) $k \leq -10$:

(1014) $x \leq 9$: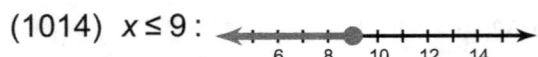

(1015) $n > 34$: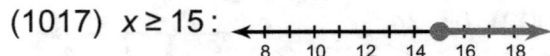

(1016) $x > 31$:

(1017) $x \geq 15$: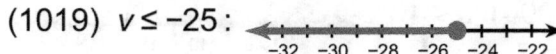

(1018) $p < -39$:

(1019) $v \leq -25$:

(1020) $x \leq 4$:

www.ingramcontent.com/pod-product-compliance
Lightning Source LLC
Chambersburg PA
CBHW081742100526
44592CB00015B/2270